Storm over a Mountain Island

Storm over a Mountain Island

Conservation Biology and the Mt. Graham Affair

edited by

CONRAD A. ISTOCK

ROBERT S. HOFFMANN

The University of Arizona Press / Tucson

The University of Arizona Press

⊗ This book is printed on acid-free, archival-quality paper.
Manufactured in the United States of America
00 99 98 97 96 95 6 5 4 3 2 1

Library of Congress Cataloging-in-Publication Data
Storm over a mountain island : conservation biology and the Mt. Graham affair / edited by Conrad A. Istock, Robert S. Hoffmann.
p. cm.
Proceedings of a workshop held in Tucson, Arizona, in October 1989, with additional papers.
Includes bibliographical references and index.
ISBN 0-8165-1551-4 (cloth). — ISBN 0-8165-1577-8 (paper)
1. Mountain ecology — Arizona — Pinaleño Mountains — Congresses. 2. Astronomical observatories — Environmental aspects — Arizona — Graham, Mount — Congresses. 3. Tamiasciurus hudsonicus — Arizona — Pinaleño Mountains — Congresses. 4. Endangered species — Arizona — Pinaleño Mountains — Congresses. 5. Wildlife conservation — Political aspects — United States — Congresses. I. Istock, Conrad A., 1936– II. Hoffmann, Robert S.
QH105.A65S86 1995 95-13667
574.5′09791′54 — dc20 CIP

British Library Cataloguing-in-Publication Data
A catalogue record for this book is available from the British Library.

Contents

PART 4 The Biogeography of the Pinaleños: *Past and Present*

PART 5 Vulnerability to Extinction of Small Isolated Populations: *Biology and the Red Squirrel*

Preface

A public controversy over plans for construction of an astrophysical observatory in the Pinaleño Mountains of southeastern Arizona swelled during 1987 and 1988. For several years the controversy attracted national attention, then for a while the interest of the press and scientific journals waned, only to revive again recently. Initially, planning for the observatory focused on a peak called Mt. Graham. The proposed location led to adoption of the name "Mt. Graham International Observatory" (MGIO), a designation that has been retained even though construction eventually occurred on nearby Emerald Peak. The original targets of several environmental groups were the University of Arizona and the Smithsonian Institution. Later, protests spread to the international partners in the project: the Max Planck Institute for Radio Astronomy in Germany and the Vatican Observatory in Italy. Several years ago Ohio State University and the universities of Texas and Pittsburgh backed away from participation in the face of protests. Recently, in 1993, controversy erupted at Michigan State University because of a proposed affiliation of that university with the MGIO; at present, Michigan State does not plan to join the consortium.

In 1987 many of the roiling environmental issues lacked clarity due to insufficient scientific information. This lack of information seemed most serious in the biological data available for the observatory's environmental impact statement. A number of people suggested that it might be possible to obtain better definition of these issues by asking a group of biologists and astronomers, with many kinds of knowledge and experience, to provide a more searching analysis of the biology of the Pinaleños, the promise of these mountains for astronomy, and ways the two branches of science might most effectively interact and respond in the public, highly charged setting that simultaneously involved issues of conservation biology and the progress of contemporary astrophysical science. In response, the University of Arizona and the Smithsonian Institution jointly sponsored a two-day workshop entitled "The Biology of Mt. Graham," held in October 1989 in Tucson, Arizona.

In addition to academic biologists and astronomers, workshop participants included biologists from the federal and state agencies responsible for the forest land and wildlife of the Pinaleños, as well as University of Arizona administrators. All sessions were open to the public. Protesters enlivened the proceedings from time to time.

The presentations formed such a stimulating and coherent body of information, analysis, and thought that we thought the proceedings should be published in a single collection. The workshop also covered scientific and logistical aspects of the proposed observatory project because it was the source of the controversy and thus raised conjoint issues of national science and environmental policies. The concerns that fueled the public controversy could not and should not be ignored. Indeed, several of the scientists providing talks at the conference (and chapters in this book), as well as many audience participants, were then and are now opposed to construction of the observatory. To produce a more complete account of the "Mt. Graham affair" we solicited additional chapters covering the flow of events and political and legal issues.

We think the story of Mt. Graham aptly fits Interior Secretary Bruce Babbitt's image of an environmental "train wreck"—a "train wreck" on our path to learning how a more civil United States of the twenty-first century might wisely and effectively address environmental problems of many kinds in many places before they become raging and expensive confrontations. Whether the Mt. Graham collision might have been avoided will be for readers to decide for themselves, but possibly, given our present state of scientific acumen and political and social wisdom, it was inevitable. The conflict the Mt. Graham affair stirs in many minds was eloquently expressed by Fred Gehlbach as he reflected on his experience at the workshop in the preface to his 1993 book *Mountain Islands and Desert Seas:*

> Is it possible to build the observatory and still have red squirrels and spotted owls—to blend culture with nature? Certainly none of us wants to lose the two creatures and the many other beasts and plants

that live with them in community. Yet I, for one, am curious about everything in the cosmos and realize the desirability of this place for telescopes as well as organisms.

We owe many debts for assistance at the workshop and afterward from the following people: Sue McQuirk, Buddy Powell, John Ratje, and Peter Strittmatter (Steward Observatory); J. T. Williams (Smithsonian Institution and Steward Observatory); Maria Ballantyne and Ross Simons (Smithsonian Institution); Michael Cusanovich (vice president, University of Arizona); and Amy Chapman Smith (acquiring editor), Sally Bennett (manuscript editor), and many others at the University of Arizona Press. Our thanks go to each of these people. We owe special thanks to Nancy L. Istock, who organized and coordinated the complex task of producing the final manuscript. Without her tireless and enormous effort the book would never have been finished. We owe her the largest debt of all.

Storm over a Mountain Island

Introduction

Occurring this very night
By no established rule,
Some event may already have hurled
Its first little No at the right
Of the laws we accept to school
Our post-diluvian world:

But the stars burn on overhead,
Unconscious of final ends,
As I walk home to bed,
Asking what judgement waits
My person, all my friends,
And these United States.

W. H. AUDEN, "A Walk After Dark"

Individual presenters and participants in the ensuing discussions during the 1989 workshop on the biology of Mt. Graham did not restrict their analyses and remarks to Mt. Graham (High Peak), a single peak in the Pinaleño Mountains. The organizers of the workshop intentionally broadened its biological purview to cover the entire high-elevation ecological system of the Pinaleño range and its larger biogeographical context within the vast archipelago of mountain islands stretching from the southern Rocky Mountains across the Great Basin and Colorado Plateau to the Sierra Madre cordillera of Mexico. Participants also explored conservation issues involving a broad view of habitats, populations, and natural communities, followed by an intense focus on the population biology, behavior, and genetics of the central player in the drama — the Mt. Graham red squirrel.

The importance of the Pinaleño Mountains from several detailed biological perspectives vividly emerged during the workshop. Revisions and expansions of these presentations form the core of the book and make, we think, an irrefutable argument that the Pinaleños hold an irreplaceable, but recoverable, ecological system. For biologists, the papers provide a rich offering of the fruits of modern biological research involving ecology, behavior, biogeography, population genetics, dendrochronology, paleobiology, extinction, evolution, natural history — and conservation biology. They provide a framework for a much deeper and more comprehensive future understanding of montane ecological systems in general, and specifically for the mountain island "archipelago" of the U.S. West and Mexico.

As we began organizing the workshop papers into this book it became clear that policy issues — struggles to define and implement policy — as well as dramatic real-world events formed essential parts of efforts to build or defeat the Mt. Graham International Observatory (now in the advanced stages of construction but facing continuing litigation). As a study in conservation biology and a guide to the crafting of future social and environmental policy, the whole story had to be included. We have tried to accomplish these goals through the inclusion of several chapters that were not part of the original workshop. These additional sections include the opening part recounting and discussing the thought-provoking saga of the Mt. Graham observatory: unfolding events, the environmental regulatory legislation and policies involved, and much of the underlying politics. We requested an essay discussing the future of ground-based astronomy from Martin Harwit because opponents of the Mt. Graham observatory frequently asserted that observatories in space would quickly make all ground-based observation unnecessary. This chapter appears in Part 2 along with the workshop paper on astronomy and the Mt. Graham observatory by Peter Strittmatter. We asked Paul Martin, who attended the workshop, to reflect on the Mt. Graham affair six years later; his thoughts appear in Chapter 15.

However, another aspect of the Pinaleño Mountains is particularly important from the perspective of general, present-day conservation practices. With a geographical area of about 60,700 ha (150,000 acres), the Pinaleños are intermediate in spatial scale between large wilderness areas and small park lands and preserves. With its dramatic vertical diversity from desert to high-elevation spruce-fir forest, this montane ecosystem is an exemplar of medium-sized natural areas close to civilization that need to be protected if North America is to retain a significant part of its remaining biological diversity.

And there is the equally general purview of astronomy, as astronomers search for knowledge and understanding about ever wider reaches and ever more diverse phenomena in the universe — a search for all the extraterrestrial objects and phenomena present in the universe and where and how they emerge and evolve. Someday, perhaps, astronomy may provide conclusive knowledge about the existence of life elsewhere in the universe. But short of knowledge of life elsewhere, astronomy continuously provides new understanding of the earth's place in the universe—the one planet on which we know life did appear and evolve, and on which we must preserve life for our own and nature's sake. Thus, the joining of astronomical and biological knowledge has more than esoteric meaning; it can bring better and more useful understanding of the conditions for life on earth. Knowing these things, the astronomers associated with the creation of the Mt. Graham observatory were chagrined to find themselves cast as rapacious destroyers of a pristine wilderness.

The title of this book is literal in two senses. Natural storms of great ferocity sweep onto Southwestern mountain islands, including the Pinaleños. These storms, however, occur infrequently and tend to clear quickly in the thin air over the Mt. Graham International Observatory (MGIO) on Emerald Peak, which exceeds 3,050 m (10,000 feet) in altitude. The many sparkling-clear days and nights between storms offer astronomers an excellent view of the universe over a wide portion of the electromagnetic spectrum. In a second and equally real sense, the ferocious opposition to the MGIO broke like a storm over the Pinaleños and spread all across this country, into the halls of government, and to Europe. Unlike its meteorological analogue, it has been slow to abate.

This book also provides a historical record of much that transpired during the Mt. Graham affair and a record, as well, of the present state of the kinds of fundamental scientific knowledge we need to assess the many public events that fostered a plethora of articles, editorials, and pamphlets, and one television documentary, as well as considerable political maneuvering. Through the pages of the book many different voices speak, sometimes expressing viewpoints that are partially or totally at variance. No attempt has been made to hide or reconcile such differences and conflicts; they are vital and instructive parts of the historical record.

The theory and practice by which a balance between development, nondestructive uses, and enduring preservation can be achieved within intermediate-sized natural areas such as the Pinaleños needs to be brought to the state of a fine scientific and political art as part of conservation biology. This goal can be achieved only through a better understanding of the complex interactions of biology, public policy, laws and their interpretation and enforcement, news media, political action, esthetics, human aspirations, and history—all starkly illustrated and complexly intertwined in the Mt. Graham story. Lessons from the Mt. Graham experience emerge in many of the papers included here, and we examine a few of these briefly at the end of the book. Undoubtedly many other lessons, insights, or conclusions have been overlooked. We hope that the publication of this story will allow others to find what we have missed and that they will share their thoughts and observations with others. Above all, we want this book to be pondered, discussed, and criticized by students, environmentalists, biologists, resource managers, reporters, lawyers, and laypeople—all who are concerned with the preservation of nature and the future sojourn of human life on earth, which is inextricably bound to the rest of the great diversity of life.

1

The Setting and the Flow of Events

Environmental Decision-Making in the Pinaleños

Standing here in the deep, brooding silence all the wilderness seems motionless, as if the work of creation were done. But in the midst of this outer steadfastness we know there is incessant motion and change. JOHN MUIR, *Wilderness Essays*

Part 1 provides three independently written overviews of the Mt. Graham saga. Together these essays set the contemporary social and political stage for the explorations of the biology and astronomy to come in later chapters.

From Washington, D.C., M. Mitchell Waldrop surveys the saga from the perspective of a science writer, physicist, and close observer of the Washington scene. His quotations of crucial statements by the administrators, scientists, resource managers, and environmentalists who were most directly involved with the saga are unmatched elsewhere. The second overview contains the recollections and conclusions of the senior editor, Conrad Istock, who observed the events and developments firsthand as a biologist at the University of Arizona and was periodically drawn into the fray. The third chapter, by Ross Simons, provides a detailed treatment of the regulatory and political activities that played prominent roles in the unfolding of the saga. Every student of environmental biology needs to be conversant with the implications of this essay, whether agreeing or not with the author's conclusions.

Although all three essays cover some of the same ground, the first is rich with its emphasis on public policy and the political undercurrent of environmental issues; the second describes more of the concerns of astronomers and conservation biologists; and the third lays out the framework of the legal restrictions within which the saga unfolded. Given the multitude of perspectives evidenced during the Mt. Graham affair, even these three viewpoints give the reader only a sample of the issues and opinions that contributed to this storm over the mountains.

CHAPTER ONE

The Long, Sad Saga of Mt. Graham

M. Mitchell Waldrop

In the annals of environmental warfare, the battle over Arizona's Mt. Graham has been as nasty as any: an exercise in recrimination and venom equal to the strife over the snail darter and the spotted owl.

The struggle, which began in the mid 1980s, reached a critical point at decade's end as the University of Arizona prepared to break ground on a $200 million astronomical observatory in the middle of a mountaintop forest that was the sole habitat of an endangered subspecies known as the Mt. Graham red squirrel. Environmental activists, claiming that the observatory would destroy the squirrel, plastered the university campus in Tucson with "No Scopes" stickers. The radical group Earth First! hinted repeatedly that the mirrors of the telescopes would be smashed if the observatory was built. In 1989, vandals cut power lines to the nearby Kitt Peak National Observatory and mailed a dead ground squirrel to the home of the university's astronomy director, Peter Strittmatter. And in February 1990, someone mailed a death threat to Arizona biologist Conrad Istock, one of the few professional biologists to support the project publicly. The letter with the death threat arrived a few days after a conservation biology seminar by Istock based on much of the material in Chapter 2. A week later a second letter of identical printing arrived; it retracted the threat and apologized.

In Washington, meanwhile, the General Accounting Office launched an investigation into allegations that the university had politically "fixed" the permitting process for the observatory. And more than a dozen national environmental groups pulled out all the stops to lobby and litigate against the project, claiming that the university had undermined the National Environmental Policy Act and the Endangered Species Act by getting a special exemption from Congress.

Natural Enemies?

Although this kind of imbroglio might seem all too familiar to researchers who have had to face animal rights demonstrations or genetic engineering protests, the astronomers generally found it baffling and hurtful. They had never before been accused of being the bad guys. Quite the opposite. Many of them are ardent environmentalists themselves—not only personally, but professionally as well: they cannot even begin to do their work unless they have a clear view of the heavens untrammeled by city lights and pollution. By rights, said Peter Boyce, executive director of the American Astronomical Society, astronomers and environmentalists ought to be natural allies: "Here are two groups that ought to be walking arm in arm with respect to protecting darkness, the night sky, and remoteness."

Much the same feeling also seemed to prevail in most of the established environmental groups. It was the Sierra Club, after all, that had published one of the best popular introductions to astronomy in recent memory: Timothy Ferris's *Galaxies*. "[The astronomers] are not the enemies we would have chosen," said Robert Smith, Sierra Club representative in Phoenix.

But if astronomers and environmentalists are natural allies, why did they spend the better part of a decade battling each other over a mountaintop in Arizona? Could this whole fight have been avoided?

The short answer is no. When one side says "Scopes" and the other side says "No Scopes," there is little room for negotiation. However, the dispute did not have to become *this* polarized. The fact is that the Mt. Graham affair was a public relations disaster for the University of Arizona, concerned environmental groups, and the involved federal agencies—a long, sad saga of naiveté, poor communication, missed opportunities, and finally, desperation.

A Golden Opportunity

This was the first time that the university had ever had to cope with a major environmental controversy. So when the astronomers' plans began to take shape in the early 1980s, neither they nor the university administrators had any feel for the sensitivities involved.

To chief astronomer Strittmatter and his team, Mt. Graham was simply a golden opportunity. Interest in building a new generation of high-technology telescopes was burgeoning worldwide—not least because of the methods being developed by University of Arizona astronomer Roger Angel to produce inexpensive telescope mirrors as much as 8 m in diameter, or 60 percent larger than the venerable 5 m instrument on Mt. Palomar in California. So if the University of Arizona could develop a new site to host some of these telescopes, Strittmatter reasoned, it stood to become a world-class power in astronomy. Mt. Graham seemed the natural choice (see Chapter 5). Although it is by no means the best astronomical site in the world—that honor goes to Mauna Kea in Hawaii and to the Andean peaks in Chile—its atmospheric quality was deemed more than adequate. It was not threatened by light pollution, as were existing facilities such as Kitt Peak National Observatory. It was not likely to be locked away in a wilderness area or national park, as were several alternative peaks. It lay within the state of Arizona, where previous observatory projects had been supported by the state's congressional delegation. And, most important for keeping the construction costs down, it was relatively accessible: Mt. Graham was only a three-hour drive from Tucson, and it already had a road to the summit.

In retrospect, it is ironic that during preliminary discussions the environmental impact of the project seemed to be a minor concern. As part of the Coronado National Forest, Mt. Graham had long since been given over to the Forest Service's multiple-use philosophy, which meant vacation cabins, logging, and hunting—including squirrel hunting. If anything, the astronomers felt that they would be helping preserve the mountain by setting aside the summit as an astrophysical area with sharply restricted public access.

A Breakdown of Communications

What Strittmatter and his fellow astronomers were not bargaining for, however, was the involvement of a cadre of environmental activists who were incensed by the Reagan administration's pro-development stance and by the unrelenting pressures of development in the burgeoning Sun Belt. "There are people who just feel enough is enough!" said Robert Tippeconic, who was chief forester for the Coronado National Forest during most of the controversy. To those people, the astronomers looked suspiciously like developers trying to lock up the last high mountain in southern Arizona that did not have telescopes on it already.

Could the conflict have been headed off at this point? Possibly. Strittmatter claimed that he did try. In 1984 he put out feelers to several environmentalist and citizens' groups to join an external advisory committee on the project, with the express purpose of maintaining a dialogue. There was even a fair amount of interest, he recalled.

However, administrators of the Coronado National Forest, the managing agency, would have had to approve any such committee, and chief forester Tippeconic opposed it. Tippeconic asserted that the Forest Service could handle its own public relations. Moreover, the National Environmental Policy Act (NEPA) required opportunity for public comment throughout the preparation of an environmental impact statement. And in any case, he felt that a committee put together by Strittmatter would be widely perceived as being a rubber-stamp body, stacked with environmentalists handpicked by the university. "The Forest Service opens its doors to everyone," Tippeconic said.

So Strittmatter let the matter drop—a decision he said he came to regret profoundly. Not long afterward, he and his colleagues learned that there is a world of difference between having a dialogue with people during a project's design, when concerns can be worked out quietly, and letting people see it only during the NEPA public comment period *after* it is designed—when it tends to look like a fait accompli.

In 1984 the astronomers were told by the Forest Service that the environmental impact statement would require a description of the maximum size their project could ever possibly be. So, as the astronomers tell it, they innocently drew up a site design that crammed a telescope onto every spot that could possibly take one. Astronomers proposed 18 telescopes, far more than anyone had definite plans for. But to the public, of course, the site plan looked like "This is what we're going to build." The environmentalists were outraged: to them it looked like a total rape of the mountaintop. The opposition, which to that point had been confined to a few committed individuals, began to grow.

An Alternative That Fizzled

Still, by 1986, Strittmatter's vision of Mt. Graham as a new world center of astronomy seemed to be coming true beyond all expectations. The Smithsonian Astrophysical Observatory, the Max Planck Institute in Germany, the Vatican Observatory—all were either interested or committed. The University of Arizona was forming a consortium with several other institutions to build the Columbus telescope, which would have two of Angel's 8 m mirrors arranged like a pair of binoculars. Indeed, the Mt. Graham project had far outgrown Strittmatter's department; overall authority for politics and fund-raising on the project was vested in Laurel Wilkening, who was at that time vice president for research at the University of Arizona.

By this point, however, every university official involved with the project also recognized that they had to get serious about the environmental issues. Biological surveys conducted for the draft environmental impact statement (which was not released for public comment until October 1986) had underscored the fact that the spruce-fir forest on the summit had been an isolated "sky island" for some 8,000 to 11,000 years, since the last ice age. "In just six weeks of research we found six new species of insects," said survey leader Peter Warshall of the university's Office of Arid Lands Studies. The biologists also identified several unique species or subspecies of plants, snails, and rodents. And, of course, the university had to take into account the Mt. Graham red squirrel, whose estimated population of 328—considered dangerously low by Warshall and other wildlife biologists—made it a prime candidate for listing under the Endangered Species Act.

Originally, at least, Wilkening and her colleagues had announced that they would not oppose the listing. But then, in an effort to figure out just how big a complication the listing was going to be, Wilkening contacted California attorney Robert Thornton, who had formerly worked on the Endangered Species Act as a staff counsel on Capitol Hill. "He advised us that listing the red squirrel and going through the entire process would likely tie the University up in litigation for a long time," she said. However, Thornton suggested an alternative: protect the squirrel and its habitat so thoroughly that it would not need to be listed. (The Endangered Species Act allows the secretary of the interior to consider existing conservation plans when deciding upon a listing.) He suggested that the university offer to institute a habitat conservation plan immediately, involving intense study and conservation of the entire mountaintop ecosystem, and that it establish an ongoing dialogue with environmentalists and citizens' groups to monitor the plan while the observatory was under construction.

Wilkening and her colleagues loved the solution. "It seemed like a logical, rational plan to achieve our goals *and* to preserve the habitat for the squirrel," she recalled. Confident that they would be greeted as heroes, she and Thornton unveiled the plan in a public hearing on 26 August 1986. "We are proposing an alternative to the listing of the Red Squirrel as endangered," she began.

It was not what the activists wanted to hear. The next day, the headline in the *Arizona Daily Star* read "UA asks U.S. to drop rare squirrel from endangered list," and outraged opponents of the project were already dismissing Thornton's proposal as a hash of bizarre and cynical legalisms. The real message, they were convinced, was that the university planned to fight the listing tooth and nail.

"We really thought we were trying to do the right thing," said Wilkening. "But to present the plan at a public hearing and expect people to say 'Oh, how wonderful,' was naive." Apparently it had never occurred to anyone to talk about the plan with the opposition groups beforehand.

In any case, the issue was soon moot. Tippeconic's response to the university's proposal was that the Forest Service's land use plan would protect the squirrel sufficiently. The red squirrel was duly listed as an endangered species on 3 June 1987 (Chapter 3), and the protests started to become venomous.

A Desperate Look to Congress

By the summer of 1988, the university's Mt. Graham team was frantic. After two years of complex negotiations with the Forest Service, members of a university task force had finally come up with a site plan that they asserted would minimize the environmental impact to the mountaintop and that seemed acceptable under the NEPA guidelines. In the process they had reduced the number of telescopes from 18 to the currently planned maximum of seven, with the total affected area—including a buffer zone as well as the actual construction sites—to be no more than 10–15 ha (24–38 acres) out of a total squirrel habitat of some 4,450 ha (11,000 acres). As astronomer Roger Angel declared in exasperation, "The environmentalists *won*!"

Yet the protests kept getting louder, with demonstrators roaming the campus in squirrel suits and Earth First! talking eco-sabotage of the telescopes: to the activists, seven telescopes were still seven too many.

To Strittmatter, the delays were not just annoying; they were becoming dangerous. His coalitions were not going to hold together forever. The University of Texas had already pulled out, and the Germans were starting to grumble. The final straw came on 14 July 1988. As one of the last steps in the environmental review process, the U.S. Fish and Wildlife Service formally issued a biological opinion containing three "reasonable and prudent alternatives" under which the university could build its telescopes without an intolerable jeopardy to the red squirrel. The third alternative in that list was one that Strittmatter and the other astronomers thought they could accept: first the university would close and reforest the existing road to the summit, which mostly runs through prime squirrel habitat; then it would cut a much shorter road, running mostly through mediocre squirrel habitat, to a wide crag of the mountain known as Emerald Peak. There they could build three telescopes, with the other four possibly to come later once biologists had determined the impact of the first three on the squirrel population.

Despite this progress toward a solution, however, chief forester Tippeconic still considered himself required by NEPA to give the public yet another opportunity for input. He ruled that because the road was something new in the observatory plan, NEPA would require a new round of public comment. Strittmatter and his colleagues erupted. Officially, the comment period was 60 days. But unofficially, he claimed, Forest Service insiders told the university that a new round of public comment would likely result in another two to four years of litigation and appeals. "And what certainty did we have that there wouldn't be another four years after that?" he exclaimed.

To Michael Cusanovich, who succeeded Wilkening as the university's vice president for research in August 1988, there was only one course to take: go to Congress, where the Arizona delegation had long since declared its support for the observatory. "Either we could exercise our constitutional rights," he said, "or we could cancel the project." In Washington, he hired the high-priced lobbying firm of Patton, Boggs, and Blow to ask Congress to exempt the project from any further requirements under NEPA and the Endangered Species Act. Accordingly, Senator Dennis DeConcini (D-Ariz.) added language authorizing the observatory to a collection of miscellaneous land-use measures known as the Arizona-Idaho Conservation Act. After a delay occasioned by House interior committee chairman Morris K. Udall (D-Ariz.), who insisted the measure be redrafted to explicitly rule out any exemption from the Endangered Species Act, the bill was passed and signed into law in November 1988.

Back in Tucson, the Mt. Graham team was exultant. "At the very least, we kept the project going," said Strittmatter. But virtually everyone else I contacted at the time considered it a Pyrrhic victory. Many of the University of Arizona faculty were outraged. "The university was saying, 'If you don't like a law, buy yourself a new one,' " said biologist Warshall.

Meanwhile, what had largely been a local environmental issue became a national issue overnight: the Arizona-Idaho Conservation Act, whatever its disclaimers, was widely viewed by environmentalists as a horrible precedent undermining the Endangered Species Act itself. In Tucson, a whole new wave of activists entered the fray. In Washington, lobbyists for the major environmental groups started pounding down congressional doors. "This is the fundamental test of whether the environmental laws of the last 20 years are worth the paper they are written on," declared Nancy Wallace, a lobbyist for the Sierra Club's Washington office.

Meanwhile, a consortium of many of those same groups, with the Sierra Club Legal Defense Fund in the lead, filed suit in federal court to force the Fish and Wildlife Service to redo its biological opinion on the grounds that the red squirrel population had drastically declined since 1988 and was then as low as 139 individuals. (The position held by the Forest Service and the Fish and Wildlife Service was that population fluctuations of this magnitude are within the range anticipated by the biological opinion.)

Meanwhile, as if the legal battles were insufficient, the political disputes continued to escalate. The General Accounting Office launched an investigation into statements by two Fish and Wildlife Service biologists that they had been ordered by their superiors—allegedly under political pressure from the university through the Arizona congressional delegation—to write "reasonable and prudent" alternatives that would allow the observatory to be built, even though they did not believe those alternatives were valid. The General Accounting Office gave its report on 26 June 1990 in a hearing before the House subcommittees on fisheries and wildlife and national parks and public lands (see Chapter 3). The critics, said the investigators, had a point: the "reasonable and prudent" alternative being followed by the university had indeed been drawn up at the behest of Michael Spear, the Southwest regional director of the Fish and Wildlife Service, and was not supported by the field studies on which the agency had based its biological opinion. Moreover, Spear had improperly used a nonbiological criterion in reaching his decision—namely, the perceived worth of the observatory. Thus, said the General Accounting Office, the opinion needed to be updated.

Not surprisingly, the environmentalists hailed the report as a "scathing indictment" of the observatory, while Strittmatter called it "an appalling piece of work." The Fish and Wildlife Service conducted yet another study, convening a five-member panel—including the two dissident biologists—to do a 30-day review of the original biological opinion. The panel concluded that the opinion probably needed to be redone, since, among other things, it had not included a consideration of global warming. Nevertheless, the university prevailed in the end. Administrators of the Fish and Wildlife Service, after conferring with the Justice Department over the nuances of the Arizona-Idaho Conservation Act, decided that the act left them little room in which to maneuver. Rather anticlimactically, the federal resource managers gave the university an official go-ahead in autumn 1990, and the university immediately started clearing ground.

And the red squirrel? The impact to date seems to have been minimal. Recent studies indicate that the squirrel can actually function quite well in the mixed-conifer forest lower down on the mountain, and the monitoring program thus far has found that the squirrel population in the construction areas is faring as well as that in untouched areas (Chapter 14). University foresters have also been digging up all the young trees (18 inches or smaller) in the cleared areas and using them to reforest open areas left by previous clear-cutting. Looking to the longer term, moreover, the university's agreement with the Forest Service calls for a ten-year program of squirrel research and habitat conservation on the mountain, at the rate of $200,000 per year.

Ultimately, however, some of the squirrel's habitat will be lost, which means that the squirrel will suffer some marginally reduced chance of survival—although precisely how big a reduction is a matter of endless debate. "The squirrel will not die from the telescopes alone," said biologist Warshall. "The squirrel will die because of a catastrophe such as a fire or a spruce budworm infestation"—with the habitat lost to the observatory perhaps making the critical difference (see Chapter 11).

Either way, the political legacy of bitterness and mistrust is still there, and the long-term cost to the university (and, for that matter, to the environmental groups and the federal resource management agencies) has yet to be reckoned. The Mt. Graham affair seems destined to go on record as one environmental conflict that passed beyond all reach of goodwill and reason—unnecessarily. "People will look back," said chief forester Tippeconic ruefully, "and cite this as one of the classics of the genre."

Acknowledgments

This essay has been revised from a report that appeared in the journal *Science* on 22 June 1990.

References

All direct quotations were obtained by personal interviews conducted by the author. All other assertions are derived from these interviews, together with documentation supplied by the University of Arizona, the U.S. Forest Service, and other involved groups. The notes for the interviews are archived at the offices of the American Association for the Advancement of Science.

CHAPTER TWO

Telescopes, Red Squirrels, Congress, Courtrooms, and Conservation

Conrad A. Istock

The prospects and the trouble began well over a decade ago. In 1983 the Steward Observatory of the University of Arizona and the Smithsonian Institution, and later in collaboration with other U.S. universities, the Vatican Observatory, and the Max Planck Institute for Radio Astronomy in Germany, proposed that a new generation of ground-based, astronomical instruments be constructed in the Pinaleño Mountains of southeastern Arizona, a part of the Coronado National Forest. Later, financial support for the project came also from the U.S. National Science Foundation. This was a major scientific project of great importance for national and international astronomical research. It was "big science," too, entailing visions of considerable prestige and money. The astrophysical complex was called the Mt. Graham International Observatory (MGIO) because one of the peaks in the Pinaleños thought by astronomers to be a superb natural platform for telescopes was known as Mt. Graham. Although another peak above 10,000 feet (3,050 m) in elevation, Emerald Peak, subsequently became the site of astrophysical development, MGIO remained the official name of the project.

Progress toward production of the environmental impact statement required under the National Environmental Policy Act proceeded for over half a decade, through numerous, often tense, discussions involving the University of Arizona, the U.S. Forest Service, the U.S. Fish and Wildlife Service, and the Arizona Game and Fish Department. By the middle of the 1980s rumblings of objection from individuals and local groups raising environmental concerns about the MGIO had begun (especially from the newly formed Coalition to Preserve Mt. Graham), and in 1988 these rumblings exploded into a storm of unyielding protest. Earth First! seized center stage, being the most vocal and visible organization, and the Maricopa Audubon Society became the most vigorous in producing mailings and eliciting newspaper articles and editorials opposing the MGIO. Demonstrations occurred in Tucson, Phoenix, and Washington, D.C. University officials were bewildered by the vehemence and tenacity of the opposition. (Later, Ohio State University withdrew from the project, and still later the University of Pittsburgh and Michigan State University contemplated participation but withdrew in the face of local protests.)

University people generally view their institutions as sources of goodness and enlightenment and are deeply disturbed when portrayed publicly as ruthless developers. The university officials and astronomers faced a breed of environmentalists who were well organized, had an almost messianic view of their goals, and felt threatened by the prevailing anti-environmental posture of the Reagan administration. Complex hostilities that universities and other intellectual institutions (scientific in particular) frequently elicit nowadays may also have motivated some among the opposition. Soon a litigious front opened as the Sierra Club Legal Defense Fund sought injunctions to kill the project. The university retained some of the best lawyers and added many new voices of its own to the rising cacophony: some pro, some con, some "official," and many unofficial. And so the battle was joined.

The Red Squirrel as an Avatar

Early on, those assembling biological information for the environmental impact statement considered a list of species that might be affected by forest removal at the observatory site or by other construction activities. This list included among others the Apache trout, peregrine falcon, Mexican spotted owl, northern goshawk, long-tailed vole, wild turkey, monkey-face grasshopper, several wildflowers, and the Mt. Graham red squirrel (*Tamiasciurus hudsonicus grahamensis*), one of approximately 25 subspecies of this typically common and widespread animal. At the U.S. Forest Service, attention settled on the red squirrel, and in 1987 the population in the Pinaleños was formally listed as endangered under the U.S. Endangered Species Act. Some advocates of the observatory claimed that federal and state agencies intentionally assigned the red squirrel a role analogous to that played by the snail darter during opposition to construction of the Tellico Dam years ago; that is, it was made a symbol around which to rally opposition even though no one had determined whether the observatory would threaten its existence.

The population of red squirrels in the Pinaleños is one of only a few scattered populations along the southwestern boundary of the species range and has possibly been isolated from all conspecific populations for 8,000 years or longer (see Chapters 12 and 13). The population is small, probably less than a thousand individuals at times of maximum size, and may be close to extinction due to natural postglacial warming, logging, and wildfire burning over much of its habitat. It is the kind of population that elicits intense interest from ecologists, evolutionists, and conservationists (see Chapters 10–14). Symbol or not, this population is certainly an avatar representing a profoundly human and urgent issue—the growing desire of many people to preserve nature on a much wider scale and with much greater assiduousness.

We have an opportunity to learn much from this avatar, for the storm over the MGIO is not an isolated cause célèbre. Granted, this controversy has unique geographical and biological characteristics and involves a somewhat unusual amalgam of environmental and science policy issues, yet it is similar in many respects to other struggles between preservation and exploitation such as the cutting of north-temperate old-growth forests, global species extinction and the maintenance of biological diversity, rain forest harvesting, wetlands destruction, and petroleum exploration and extraction in wildlife reserves and oceans. Measured only in the tens of thousands of hectares, the Pinaleños nevertheless represent a recoverable ecological system of great biological interest. Though dangerously beset with past human interventions resulting from the multiple-use policy of the Forest Service, this mountain range, and its montane archipelago, can yet become an enduring redoubt for part of the wonderful biological diversity of the Southwestern mountain islands of the United States.

A Long and Tortuous Path to Congress

A remarkable part of the Mt. Graham story involves the multiple "personalities" and roles of the federal and state agencies and the people officially responsible for the protection and use of the Pinaleño Mountains. Before and since the public storm, the stance of many in the Forest Service, the Fish and Wildlife Service, and the Arizona Game and Fish Department has often been adversarial—and sometimes subtly, sometimes openly, opposed to the observatory. Part of this opposition was due no doubt to the fact that the initial plans for the observatory suggested that many hectares of timber, perhaps 40 or more, might be cut, with two or three different peaks eventually occupied by up to 18 telescopes. Feelings that the observatory challenged the traditional responsibility of the agencies for the management of national forest lands and wildlife may also have been a factor. In addition, "outsiders" (including this author) criticized the agencies during many internal discussions and public presentations for their past management practices, especially the multiple-use policy, which had had destructive consequences. Not surprisingly, agency biologists were sensitive to these charges and often reacted defensively, though at times they also appeared to admit chagrin over the results of the multiple-use policy in the Pinaleños.

To appreciate the fascinating twistings and turnings, and the ironies, experienced by these agencies and the university, it is necessary to examine the situation in the Pinaleños prior to and after the explosion of public controversy. In the early 1980s, as now, the destructiveness of past management practices in the Pinaleños was apparent. Almost everywhere there were signs of disturbance, exploitation, and damage caused by humans. Timber harvesting and road construction in many places had destroyed or damaged large areas of native vegetation. The lush ground vegetation in the wet meadows of the high-elevation cienegas was periodically churned up by horses or pack animals or was trampled by large groups of campers. Several developed campgrounds existed, many private cabins had been built, a Bible camp installed, and a Forest Service work camp constructed. Old logging roads were apparent, and larger, more permanent roads were maintained through the mostly uncut spruce-fir forests to reach some of the highest peaks such as Mt. Graham (High Peak), Hawk Peak, and Emerald Peak. An artificial lake flanked by a campground had been created. An electronic transmission site had been installed on one peak. Young conifers were annually cut and sold as Christmas trees. Fuelwood harvesting, off-road vehicles, and snowmobiles were permitted. The red squirrels along with other game animals were hunted.

Despite the damage, the higher-elevation forests of this mountain island were still incredibly beautiful, reasonably intact in many places, and instantly fascinating to a naturalist. Analysis of a core from the High Water (Emerald Springs) Cienega demonstrated that the spruce-fir forest at the top of the Pinaleños is truly an ancient forest—a relict Holocene plant community (see Chapters 6 and 7; Anderson and Shafer 1991). The dominant tree species are Engelmann spruce and corkbark fir. The older, second-growth, mixed-conifer forests cut many decades earlier had advanced furthest in recovery and were exceptionally pleasant, though they were slated for a second harvesting during the 1980s. The recovering second-growth forests were interesting in another respect, being ideal for future studies of ecosystem recovery. A fantastic opportunity existed for the creation of a conservation plan by biologists from the agencies and the university in connection with—indeed, encouraged by—the MGIO. Such a plan did indeed emerge, at least in part, but not by such a direct and rational process.

In fact, in the 1980s the Forest Service intended to continue all former multiple-use practices and to expand camping and recreational facilities (USFS 1986:Table 14). In seeming contradiction, the Forest Service also proposed that almost 62,000 acres (25,000 ha) in the Mt. Graham area become a new wilderness area (USFS 1986:79). Simultaneously, the Arizona Game and Fish Department recommended federal listing of the Mt. Graham red squirrel as an endangered organism and shortly thereafter banned its hunting; this was in 1986, ten years after the agency had added the Mt. Graham red squirrel to its own listing of threatened wildlife in Arizona, but during which decade no protective measures were initiated (Coronado National Forest 1988:10–11). After the Mt. Graham red squirrel was proposed for federal listing, statements made by university officials (especially the vice presidents for research, first Laurel Wilkening and later Michael Cusanovich) were easily and correctly interpreted as university opposition to listing the population as endangered.

Biologists from the Arizona Game and Fish Department undertook a brief population study of the Mt. Graham red squirrel in 1985 after the University of Arizona coalition proposed the observatory project. In autumn 1986 several agencies and the university cooperatively produced the expanded biological assessment (Coronado National Forest 1988:esp. Fig. 5). The expanded biological assessment relied heavily on an initial survey of red squirrel middens (a midden is a large food cache of unopened cones buried by one squirrel in a pile of conifer cone scales [see Smith and Mannan 1994]). A great deal of discussion about matters of conservation and land use ensued as the Fish and Wildlife Service prepared the final biological opinion on the red squirrel population, as required under the Endangered Species Act. The biological opinion, released in July 1988, proposed among its other alternatives that three telescopes be built at Emerald Peak including construction of a new access road, all to be confined to 8.6 acres, or 3.5 ha (Spear 1988). It also states that requests for "additional development . . . would be accommodated" under provisions of the Endangered Species Act (Spear 1988).

The biological opinion contains a map (Spear 1988:Fig. 3) showing how biologists thought the red squirrel middens were distributed in the Pinaleños, with the proposed access road to the observatory shown on it. The fact that the new access road and observatory site appeared to lie in an area of few middens, thus minimizing impact, made sense at the time. However, later monitoring activities revealed that many more active red squirrel middens existed near the proposed road than were shown on the map (Chapter 14).

In terms of red squirrel protection the proposed solution was especially appealing because it further required that existing roads of the high country be removed; thus the taking of 8.6 acres (3.5 ha) for the two-mile access road and a three-telescope observatory would be compensated for by the return of some 60 to 80 acres (24–32 ha) when existing roads were obliterated and revegetated. The roads to be removed ran up to and right along the spine of the highest country of the Pinaleños and through areas of relatively high midden density in otherwise virgin spruce-fir forest. Removing the roads would be a crucial and dramatic step toward restoring the wildness of the high country. Other constructive preservation and restoration measures were also required in the biological opinion; these included transplanting trees from construction sites, conducting studies of red squirrel life history, monitoring the red squirrel population, and reforestation (Spear 1988: 34–35).

The "reasonable and prudent alternative" of the biological opinion that recommended placing three telescopes on Emerald Peak was the result of negotiation between the agencies and the university. The astronomers were forced to define a scaled-down project that would still create an effective or viable observatory. The project also was required to adhere to far more careful and rigorous construction and operational practices than had been required of any previous observatory, since this observatory was to be part of a sensitive and valuable natural ecosystem (see Chapter 15).

The biological opinion was signed by Michael Spear in his capacity as regional director of the U.S. Fish and Wildlife Service. Sensible as the plan of his opinion seemed to some, acceptance of the plan had not been the intent of its principal author within the Fish and Wildlife Service. The author, staff biologist Lesley Fitzpatrick, stated to investigators for the Government Accounting Office that she had fashioned a "poison pill" that would make the observatory project simultaneously unacceptable to her own agency, the Forest Service, the university, and the public (Fitzpatrick 1990a). In a deposition she said, "I am opposed to the project going on to Mt. Graham. I have held that opinion ever since I first heard of the project" (Fitzpatrick 1990b). Eventually, during a congressional hearing, she turned on Director Spear, claiming he had inappropriately ordered her to write an opinion including the three-telescope option for Emerald Peak (Fitzpatrick 1990c).

At the same hearing, Spear testified that he had done precisely what the Endangered Species Act required in this case by protecting the red squirrel population and that he was not willing to expand the interpretation and implementation of the act to other purposes (see Chapter 3). Some press reports immediately claimed the biological opinion had been "fudged." This claim was misleading, since Spear as director had final responsibility for the opinion and its contents: he alone was accountable for it, and the document must render his opinion.

Ironically, Fitzpatrick's "poison pill" provided a worthy nucleus for a conservation plan. And with the issuance of the biological opinion in July 1988 all requirements under the Endangered Species Act were fulfilled except for delineation of critical habitat for the red squirrel population; the Fish and Wildlife Service fulfilled the latter requirement in January 1990 by establishing a 1,700 acre (690 ha) red squirrel refugium.

Immediately upon the appearance of the biological opinion, Robert Tippeconic (then forest supervisor for the Coronado National Forest) and other Forest Service officials said that delays of months or a year, or even more, must ensue to allow review of its new items, especially the substitute access road. Moreover, Tippeconic said the final environmental impact statement could not be completed until after lengthy consultation and renewed opportunity for public comment. Many such delays had occurred previously, but this one was not to happen.

About six years had passed since the astronomers had requested that the administrators of the Coronado National Forest consider the Pinaleños for a future astronomical site. Frustration on the university side soared at the prospect of yet another long delay. Arizona senator Dennis DeConcini swiftly attached a congressional amendment authorizing an observatory to the Arizona-Idaho Conservation Act, which was then before Congress. The amended act passed in October 1988. Only a few weeks later, in early November, the Forest Service issued the printed and bound final environmental impact statement (USFS 1988), and thus the preconstruction requirements of the National Environmental Policy Act were satisfied.

The preferred alternative of the Forest Service, somewhat surprisingly, included a possible second phase of development to seven telescopes on 24 acres (10 ha) at Emerald Peak (something the Forest Service had refused to discuss previously), as had the amendment just passed by Congress. It appears the Forest Service followed the congressional lead precisely in shaping its final environmental impact statement. In late November 1988, President Ronald Reagan signed the Arizona-Idaho Conservation Act into law; the observatory had its mandate. Opponents claimed the congressional action was an "end run" around the Endangered Species Act and the National Environmental Policy Act—a disastrous step toward the destruction of these central instruments of environmental protection (Warshall 1994).

Implementing a Conservation Plan After All

For many people following the saga of the MGIO, all subsequent events were rather predictable ones of legal maneuvering, hearings, arguments, sensational news coverage, sober editorials, protest demonstrations, and wellsprings of propaganda on all sides. During 1989 and 1990 few realized that the denouement of a remarkable process in real-world conservation biology was quietly unfolding.

The Forest Service assigned the task of establishing an extensive red squirrel study and monitoring program to the university because the agency had no such capability and also required that the fieldwork be funded (approximately $200,000 per year) by the university, as it has been ever since. A multiagency red squirrel recovery team was formed. Funding for additional studies was provided by the university, the agencies, and the Research Corporation (a private research foundation based in Tucson). After peer review by scientists outside the university and modification of the monitoring plan, the recovery team implemented the plan in spring 1989; a steady flow of data began, with frequent reports to the Forest Service (see Chapter 14). Forest Service and university biologists also began to estimate the total size of the red squirrel population by working together, quite uneasily at times, revisiting previous middens twice a year and classifying them as active or inactive. These estimates were initially flawed by inadequate knowledge of the squirrel's biology and the process of midden turnover (see Chapter 14).

During the summer of 1989 biologists from Arizona State University began a study of forest habitat structure (Chapter 6). Other researchers initiated a study of the relationship between middens and surrounding vegetation (Smith and Mannan 1994). More recent biological research includes studies of microclimate in relation to midden position and forest edges (Chapter 14), an excellent survey of the composition of herbaceous vegetation (McLaughlin 1994), and a reconstruction of the fire history of the Pinaleños using tree-ring data (Chapter 7).

Data from the red squirrel monitoring program have contradicted a central tenet of earlier documents produced by the Forest Service and Fish and Wildlife Service: that the red squirrel requires virgin high-elevation spruce-fir forest habitat. The expanded biological assessment, based on earlier studies by Barry Spicer and Peter Warshall, said that "the spruce/fir association is the vegetation type contributing most of the excellent food habitat" for the red squirrel (Coronado National Forest 1988:Table 2 and p. 15) and that the density of red squirrels is greatest in the 472 acres (191 ha) of spruce-fir habitat found above 10,000 feet, or 3,050 m (Coronado National Forest 1988:Table 12). The monitoring results have subsequently shown that portions of the mixed-conifer habitats at lower elevation (9,000 feet, or 2,700 m) provide equally good, and in some years better, habitat of much greater potential expanse (Chapter 14). However, Spicer's fieldwork also provided evidence of a larger extent and variety of red squirrel habitat, but the writers of the expanded biological assessment ignored this information. Fluctuation of red squirrel densities in space and time is marked, and middens are created and disappear and go from active to inactive much more frequently than was previously thought.

No one yet knows either the total number of red squirrels in the Pinaleños or the total areal extent and relative qualities of their habitat (a small group of squirrels with middens has been found at 2,380 m [7,800 feet]), but eventually the field studies now under way will provide reliable estimates of the red squirrel population size, its year-to-year fluctuations, its spatial and temporal structure, its detailed ecological requirements and tolerances, the current extent of its habitat, prospects for natural habitat recovery (succession), and the long-term potential for persistence of the squirrel population. Good red squirrel habitat is definitely more extensive than was believed at the writing of the expanded biological assessment and the biological opinion in 1988, as a Fish and Wildlife Service team acknowledged in a July 1990 report. (Chapter 14 presents the best data and analysis for the red squirrel population available at present.)

The disagreement between conclusions in earlier official documents and newer, more extensive and carefully collected data reveals the scientific weakness likely to occur and recur given the ways federal and state agencies now produce assessments, opinions, reasonable and prudent alternatives, and impact statements. Resource managers should obtain high-quality data from intensive field studies before, as well as after, making environmental decisions. These field studies require adequate funding and responsible deployment of sufficient scientific skill. Results should be published in peer-reviewed journals and books. The present case illustrates that the time required to establish a solid scientific basis for decision need not be any greater than that consumed by current agency practices and perhaps could be considerably shorter in many cases.

Agency biologists, Earth First!, the Maricopa Audubon Society, and others often implied or asserted outright that the clearing of 8.6 acres (3.5 ha) initially, or eventually 24 acres (9.7 ha), would drive the red squirrel population of the Pinaleños to extinction, even though habitat for only one to six squirrels was to be removed. Such simple assertions should be weighed carefully even when they seem to be grounded in ecological theory (Kareiva 1994).

Does any population-demographic model or theory predict extinction of the red squirrel population due to this small intrusion? A recent population model does indicate that threshold effects followed by "time delayed, but deterministic extinction of a dominant competitor" can occur (Tilman et al. 1994; see also Chapter 11). Even populations that seem to be growing rapidly on average may still go extinct through "runs of bad luck" (Mangel and Tier 1994).

However, in the case of the Mt. Graham red squirrel, removing 11,000 acres (4,450 ha) of its habitat in the past (Coronado National Forest 1988:Table 2 and p. 20) did not bring extinction, nor did the introduction of the larger Abert's squirrel, a potential competitor, to the Pinaleños in the 1940s by the Arizona Game and Fish Department. Likewise, the construction of the access road on 5 acres (2 ha) with new clearing of about 2.5 acres (1 ha) has had no immediate detrimental effect on the adjacent squirrel population (Chapter 14). Substantial parts of the red squirrel habitat are now recovering due to cessation of timber harvesting.

Hence, there does not seem to be a "threshold risk" in the case of red squirrels in the Pinaleños. To retain credibility in matters of conservation, we need to invoke the idea of threshold risks cautiously, carefully weighing the evidence for and against such a possibility. With all timber harvesting now prohibited in the Pinaleños (see below), the greatest threat to the red squirrel population is probably habitat destruction by wildfires (see Chapter 7), but fire suppression will now be mandatory to protect the observatory, as well as the astronomers and biologists working in the Pinaleños.

Visiting the Pinaleños in 1995, we find in place an observatory designed to be compatible with a surrounding nature preserve. Aside from the immediate territory occupied by the telescopes and a support facility (and local effects created by artificial forest edges), the observatory requirements for quiet in the daytime and clear night skies without light pollution are positive ones biologically. For a considerable distance around the observatory the discharge of firearms, campfires (occlusion of the atmosphere), and nighttime vehicular traffic (headlights) are prohibited.

From time to time one hears people both in and out of the responsible federal and state agencies argue that new measures to protect the Mt. Graham red squirrel population and its habitat, and even wider reaches of the Pinaleño Mountains, are not the result of the observatory project but would have occurred anyway. This contention is not true. All of the official agency documents and the measures for protection of nature that they propose are directly attributable to the observatory. For example, in 1992 a draft recovery plan for the red squirrel population of the Pinaleños suggested a cost of $18.8 million for the ten-year recovery program (Fitzpatrick et al. 1992), the funding of which would be unlikely without the impetus of development. Unfortunately, this plan is poorly developed, and curiously, the observatory as a threat to the red squirrel population has all but faded from view in this document.

However, the most telling and impressive of the agency actions, and the ones that bestow the greatest protection upon the recovering ecological systems of the Pinaleños, are the stipulations in the 1989 Forest Service record of decision (USFS 1989). Here one reads as follows on the first page: "Amendments to the Coronado Forest Plan are described herein. These changes are the *direct result* [emphasis added] of implementing management direction for modified Alternative G." Modified Alternative G is the observatory project recommended by the Forest Service in the final environmental impact statement. In the record of decision (USFS 1989) we find the following changes.

1. "Eliminate sustained commercial timber harvest program for the Pinaleño Mountains."
2. "Eliminate fuelwood and Christmas tree harvest within suitable Mt. Graham Red Squirrel habitat."
3. "Recommend adding 442 acres [179 ha] to National Wilderness Preservation System."
4. "Delete the development of planned recreation sites in the Pinaleño Mountains."
5. "Delete High Peak (Mt. Graham) as a potential electronic site."
6. "Delay electronic site development at West Peak."
7. "Limit use of snowmobiles in the Pinaleño Mountains within habitat for the Mt. Graham Red Squirrel."
8. "Limit motorized public access to portions of the habitat for the Mt. Graham Red Squirrel."

These amendments and others in the record of decision are changes to the Coronado National Forest Land and Resource Management Plan of 1986. The most dramatic amendment is the cessation of all timber harvesting. Where in 1986 there was a continuing harvest of 200 million board feet from 122 acres (49 ha) per year, and trees were being cut, now there is no timber harvesting. Visiting the second-growth, mixed-conifer stands of the Pinaleños now, the ones so recently found hosting red squirrels, one is reminded that only a few years ago these forests were scheduled for a second cutting and that the projected rate of harvesting from the whole of the Pinaleños extrapolated to 1,210 acres (490 ha) over the next decade and 12,100 acres (4,900 ha) over the next century, almost exactly what was harvested during the past century. The record of decision (USFS 1989:Table 14) also tells us that 175 million board feet from 109 acres (44 ha) may still be harvested each year from the Santa Catalina Mountains and 80 million board feet may be taken each year from the Chiricahua Mountains, nearby mountain islands of the same "archipelago" that includes the Pinaleños (Chapters 8–10). Since 1988 the Forest Service has suspended tree harvesting in the Santa Catalinas while evaluating the status of the Mexican spotted owl, but only the Pinaleños have received official, and potentially permanent, freedom from logging.

Other biological benefits have come from the observatory project. In early October 1990, old roads high in the Pinaleños were obliterated by plowing, and tree seedlings and other plants and fungi are now pushing through the surface of the loosened soil. In terms of biological conservation, the Pinaleños today have greater protection, and a more concerted effort at ecosystem restoration, than any of the other Southwestern mountain islands.

The "poison pill" of Lesley Fitzpatrick became the potion that for once swept away the incongruous and often destructive practices of the multiple-use policy and put the first semblance of a conservation plan in its place. The Pinaleños were certainly almost forgotten among the Southwestern mountain islands prior to 1983 when the Smithsonian raised the possibility of an astrophysical site; the range is seldom mentioned in popular guides to the mountain islands. Although the observatory proposal seems to have been the catalyst that produced the radical change, responsive leadership from officials of the Coronado National Forest was in the end essential for their sharp reversal of "management direction." The process was unfortunately replete with confusion.

Another example of confusion involves the ways in which the provisions of the Endangered Species Act were interpreted and implemented. Technically speaking, the final solution with all of its countervailing conservation measures enacted in return for approval of the observatory is exactly like a "Section 10" solution under the Endangered Species Act, although representatives of the Fish and Wildlife Service throughout insisted they were approaching the case as a "Section 7" process, which does not require such conservation measures (Chapter 3).

Into the Courtrooms

The administrative decisions recounted above and the legal maneuvers that followed forced the university, the Forest Service, and the Fish and Wildlife Service onto the same side. Opposite these strange bedfellows was the Sierra Club Legal Defense Fund and others, suing to stop the observatory project and nullify the enabling legislation.

The lawsuits contained both biological objections (red squirrel extinction) and nonbiological objections (inappropriate process under the Endangered Species Act and the National Environmental Protection Act) to the telescopes. Esthetic concerns may also have motivated some of the litigation. The Emerald Peak location will conceal the telescopes from most vantage points until one is quite close to them; however, just knowing they are there in the midst of the forest and mountain crags offends some people. Ethnic or religious concerns of Native Americans also have been argued in court.

From 1989 on, supporters and opponents of the observatory have been in local and federal courts many times. Sometimes the courts ruled in favor of starting construction only to have a later decision suspend construction, and sometimes the rulings suspending construction were overturned by later decisions. Despite these fits and starts, construction of two telescopes began in 1990.

In October 1991 U.S. district court judge Alfredo Marquez in Tucson ruled in favor of the observatory (*Mt. Graham Red Squirrel, et al. v. Clayton Yeutter*). As the October 1991 court decision became known, renewed opposition mounted within the university, coupled this time, as occasionally in the past, with pleas that were ethnic and religious—for the high reaches of the Pinaleños may hold sacred places for the Apache or Hopi Indians, just as the Kitt Peak National Observatory site west of Tucson is sacred for the Tohono O'odham people. Both opponents and proponents of the observatory have sent delegations of Apaches to Germany and Italy in attempts to influence the European partners in the observatory coalition.

On 11 December 1991 the U.S. Court of Appeals for the Ninth Circuit denied another appeal by the Sierra Club Legal Defense Fund (*Mt. Graham Red Squirrel, et al. v. Edward R. Madigan, et al.*), thus approving continued construction of telescopes as originally mandated by the 1988 congressional action. In May 1992 U.S. district court judge William Copple denied a motion to halt construction brought by the Apache Survival Coalition (*Apache Survival Coalition, et al. v. The United States of America, et al.*). Legal appeals for "no scopes" are still under way, but it is now certain that "scopes" will search the heavens from high in the Pinaleños.

Construction of two telescopes was completed in 1994, and site preparation for a third telescope was also under way only to be halted by an injunction (*Mt. Graham Coalition, et al. v. Jack Ward Thomas, et al.*) in July 1994, which was sustained by the Ninth Court of Appeals in February 1995. It now appears that the Sturm und Drang, the long and fierce storm, is slowly abating.

Coda

There are no simple, universal solutions to the problems posed by our need to preserve nature and our competing desire to use natural areas for uniquely human purposes. For those who perceive the desperate importance of the present worldwide struggle for the preservation of nature, preservation takes precedence over exploitation. Though the preservationist view is easily justified scientifically and socially, despite protestations to the opposite by many people including the recent administrations of Presidents Reagan and Bush and many legislators in the current U.S. Congress, it does not automatically—by some clear, infallible formula—present us with the best course of action in any given real-world situation. As we have found in the present example, that result comes much harder, in ways and embodiments, with dimensions and complexities, that we cannot easily anticipate and control. E. O. Wilson eloquently captured some of the most fundamental dimensions and complexities in the following passage:

> In the end, I suspect it will come down to a decision of ethics—how we value the natural worlds in which we evolved and now, increasingly, how we regard our status as individuals. We are fundamentally mammals and free spirits who reached this high level of rationality by the perpetual creation of new options. Natural philosophy and science have brought into clear relief what might be the essential paradox of human existence. The drive toward perpetual expansion—or personal freedom—is basic to the human spirit. But to sustain it we need the most delicate, knowing stewardship of the living world which can be devised. Expansion and stewardship may appear at first to be conflicting goals, but the opposite is true. The depth of the conservation ethic will be measured by the extent to which each of the two approaches is used to reshape and reinforce the other. (1988:16)

Two things are certain. First, wise environmental decisions cannot be made without reliable and sufficient scientific understanding of the ecological systems in which these decisions have many of their most important consequences. Obtaining such understanding is an immense task given the many different natural and unnatural settings, all across the world, for which such knowledge is required. Second, as both scientists and human beings, we desire fundamental knowledge of the whole of the universe, physical and biological. Together these certainties may explain why biologists and astronomers feel many passions, some similar and some different, when they gaze upon the Pinaleño Mountains rising majestically from the Arizona desert.

References

Anderson, R. S., and D. S. Shafer. 1991. Holocene biogeography of spruce-fir forests in southeastern Arizona—Implications for the endangered Mt. Graham red squirrel. *Madrono* 38:287–95.

Apache Survival Coalition, et al. v. The United States of America, et al. U.S. District Court for the District of Arizona, Phoenix, Arizona. Case #CIV 91-1350 PHX WPC.

Coronado National Forest. 1988. *Mount Graham Red Squirrel: An Expanded Biological Assessment.* Coronado National Forest, Tucson, Ariz.

Fitzpatrick, L. A. 1990a. Signed transcript of testimony to a U.S. Government Accounting Office investigating team. Government Accounting Office, Washington, D.C.

Fitzpatrick, L. A. 1990b. Deposition 12 Jan. 1990, p. 115, lines 7–14. *Mt. Graham Red Squirrel, et al. v. Clayton Yeutter.* U.S. Federal District Court for the District of Arizona, Tucson, Arizona. Case #CIV 89-410 TUC ACM.

Fitzpatrick, L. A. 1990c. Statement before the House Merchant Marine Subcommittee on Fisheries and Wildlife Conservation and the Environment and the House Interior Subcommittee on National Parks and Public Lands.

Fitzpatrick, L. A., G. F. Froehlich, T. B. Johnson, R. A. Smith, and R. B. Spicer. 1992. Mount Graham Red Squirrel, *Tamiasciurus hudsonicus grahamensis,* Draft Recovery Plan. Region 2, U.S. Fish and Wildlife Service, Albuquerque, N.M. (Unpublished.)

Kareiva, P. 1994. Ecological theory and endangered species. *Ecology* 75:583.

Mangel, M., and C. Tier. 1994. Four facts every conservation biologist should know about persistence. *Ecology* 75:607–14.

McLaughlin, S. P. 1993. Notes on the botany of the "sky islands" region of southeastern Arizona. (Unpublished.)

McLaughlin, S. P. 1994. Additions to the flora of the Pinaleño Mountains, Arizona. (Unpublished.)

Mt. Graham Coalition, et al. v. Jack Ward Thomas, et al., and the Arizona Board of Regents. 9th Circuit Court of Appeals, San Francisco. Case #94-163324.

Mt. Graham Red Squirrel, et al. v. Edward R. Madigan, et al., and State Board of Regents, University of Arizona. 9th Circuit Court of Appeals, San Francisco. Case #954 F.2d 1441 (9th Cir. 1992). Formerly *Mt. Graham Red Squirrel, et al. v. Clayton Yeutter.*

Mt. Graham Red Squirrel, et al. v. Clayton Yeutter. 1990. U.S. Federal District Court for the District of Arizona, Tucson, Arizona. Case #CIV 89-410 TUC ACM. Later changed to *Mt. Graham Red Squirrel, et al. v. Edward R. Madigan, et al., and State Board of Regents, University of Arizona.*

Smith, A. A., and R. W. Mannan. 1994. Distinguishing characteristics of Mount Graham Red Squirrel midden sites. *Journal of Wildlife Management* 58:437–45.

Spear, M. J. 1988. *Biological Opinion.* U.S. Department of the Interior, Fish and Wildlife Service, Albuquerque, N.M.

Tilman, D., R. M. May, C. L. Lehman, and M. A. Nowak. 1994. Habitat destruction and the extinction debt. *Nature* 371:65–66.

USFS. 1986. *Coronado National Forest Land and Resource Management Plan.* U.S. Department of Agriculture, Forest Service, Southwestern Region, Tucson, Ariz.

USFS. 1988. *Final Environmental Impact Statement, Proposed Mt. Graham Astrophysical Area, Pinaleño Mountains, Coronado National Forest.* EIS #03-05-88-1. U.S. Department of Agriculture, Forest Service, Southwestern Region, Tucson, Ariz.

USFS. 1989. *Record of Decision and Forest Plan Amendments for the Proposed Mt. Graham Astrophysical Area.* U.S. Department of Agriculture, Forest Service, Southwestern Region, Tucson, Ariz.

Warshall, P. 1994. The biopolitics of the Mt. Graham red squirrel. *Conservation Biology* 8:977–88.

Wilson, E. O. 1988. The current state of biological diversity. In *Biodiversity,* ed. E. O. Wilson, pp. 3–18. National Academy Press, Washington, D.C.

Young, P., and H. R. Sanderson. Mt. Graham Red Squirrel Monitoring Program, Annual Report for 1991. University of Arizona, Tucson. (Unpublished.)

CHAPTER THREE

Mt. Graham and the Changing Political Environment

Ross Simons

The Mt. Graham controversy illustrates how environmental laws and the regulatory process function in today's society, influenced by citizens, legislators, governmental agencies, institutions, and special interest groups. It is not a pretty story for those who favor clear-cut answers, and yet it is precisely the shadows that illuminate when we examine the complex, often inchoate, mechanisms through which our national environmental laws are created and enforced.

The Mt. Graham saga is not an isolated occurrence, and in hindsight, it seems inevitable that controversy would surround the decision to develop the Pinaleño Mountains as a site for an astrophysical facility. The confrontations had their origins in an increasingly noisy national debate concerning the use of public lands in the United States. At the same time, it is not surprising that the proponents of development expected little resistance to the plan. The history of astronomy in Arizona had been an extremely positive one, and the reaction of the Forest Service and the public to expansion in this field had been favorable for more than two decades. In fact, the proposed observatory was consistent with the statutory mandate of the Forest Service, which for more than a century had called for multiple-use, sustained-yield land management. This tradition and concerns about it were long ago expressed by the first chief of the Forest Service, Gifford Pinchot:

> There are many great interests on the National Forests which sometimes conflict a little. They must all be fit into one another so that the machine runs smoothly as a whole. It is often necessary for one man to give way a little here, another a little there. But, by giving away a little at present, a great deal is given away at the end. (1907; quoted in Gippert and DeWitte 1989:8)

The observatory proposal was seemingly "business as usual" for all parties, but the rules were changing.

The idea of unbounded resources, and the intrinsic value of opening the frontier, long dominated attitudes toward the American West. Subsequent actions by citizens and government have done little to dispel the myth of limitless resources. Agencies such as the Bureau of Land Management, the Bureau of Mines, and the Forest Service were created in the nineteenth and early twentieth centuries to serve specific industries and special interest groups. Now, their functions and definitions clash with modern environmental ethics. Rapid urbanization of parts of the American West—and the consequent development of land, increasing tourism, and search for recreational opportunities—have brought growing conflicts. Traditional coalitions among ranchers, miners, and the timber industry are crumbling as the environmental movement flexes its muscles. This collapsing of old alliances has brought political leaders to an often uncomfortable reappraisal of natural resource management. The frontier mentality, the legacy of the westward expansion, takes as given the unregulated exploitation of the commons and fosters a deep distrust of powerful centralized government. Even in today's world, where we no longer have a place to move away to, it still rankles to have people in a distant capital making decisions that shape our lives.

The New Legal Framework and Its Implementation

During the course of the last 25 years, legislation has changed the legal framework for resource use in the West, as well as for the entire United States. Although regulatory agencies have been slow to implement some of these provisions, newer staff—trained not only in traditional land-use practices but also in modern biology—have increasingly pressured the leadership of these agencies into adopting the new conservation ethos and enforcing legislation that seeks to balance the needs and desires of industry with those of other constituencies.

Three principal federal laws governed the decision-making process for the Mt. Graham International Observatory (MGIO) proposal: the National Environmental Policy Act (NEPA), passed in 1969; the Endangered Species Act (ESA) of 1973; and the National Forest Management Act of 1976 (Bean 1977). (The last of these has antecedents in the Multiple-Use Sustained-Yield Act of 1960 and the Forest and Rangeland Renewal Resources Planning Act of 1974.) The Forest Service, as the managing agency of the Coronado National Forest of which Mt. Graham is a part, had principal responsibility for evaluating the

University of Arizona's proposal to build the observatory, both under the ESA because of the listing of the Mt. Graham red squirrel as endangered and under the Land and Resource Management Plan for the Coronado National Forest.

Contrary to public perception, the Pinaleño Mountains had long ago ceased to be pristine. Following traditional Forest Service practices, intensive logging (including clear-cutting) occurred over much of the mountain in the 1930s and 1940s, and timber harvesting occurred in critical squirrel habitats well into the 1980s. An asphalt road—built by felons in the 1950s and authorized by Arizona's legendary senator Carl Hayden—leads from the base of the mountain to the 10,000-foot peaks (3,050 m), and many logging roads crisscross lower elevations. The mountain also has many cottages, a Bible camp, and recreation areas providing hunting, fishing, hiking, and snowmobiling. Recreational activities in the Pinaleño Mountains now total several hundred thousand person-days annually, according to Forest Service estimates.

The Forest Service, in compliance with the National Forest Management Act and the Wilderness Act of 1964, found itself in the midst of its Roadless Area Review and Evaluation process (RARE II), begun in 1977, when the search for an astronomical site also began. Under the Wilderness Act, pieces of national forest are evaluated for designation as possible wilderness areas using a variety of criteria. Official designation generally follows a period during which national forest areas are candidates for wilderness study areas. In 1984, following the passage of the Arizona Wilderness Act championed by Senator Barry Goldwater and Representative Morris Udall of Arizona, the Pinaleño Mountains were classified as a wilderness study area (*Congressional Record* 1984).

In 1984, the Smithsonian Institution, with the active participation of the University of Arizona, petitioned the Forest Service and wrote the Arizona congressional delegation asking that the summit of Mt. Graham be excluded from the designation to allow the development of an observatory (personal communication, Dan Brocious, public information officer of the Whipple Observatory, 1990). This request was granted implicitly in the Arizona Wilderness Act; although, without benefit of an accompanying committee report only a select group of knowledgeable individuals knew about this exclusion.

In fact, to the inattentive, or to those uninformed in congressional practices and the regulatory methods of the Forest Service, quite the opposite impression may have been created. The Arizona Wilderness

Act calls for "certain lands in the Coronado National Forest, which comprise approximately 55,090 acres [22,000 ha] as generally depicted on a map entitled *Mt. Graham Wilderness Study Area* and dated February 1984, [to be] retained by the U.S. Forest Service, Washington, D.C." Only after examining the map does one recognize that the summit has been placed outside this wilderness study designation.

At the time, this exclusion did not raise concerns on the part of any constituency. The Forest Service, particularly the Coronado National Forest staff, had long cooperated with the Smithsonian Institution, dating back to the latter's initial survey for an observatory site in southern Arizona in the late 1960s. Then, and into the early 1970s, construction took place under Forest Service permit atop Mt. Hopkins in the Santa Rita Mountains, which were also part of the Coronado National Forest (personal communication, J. T. Williams, Smithsonian Institution and Steward Observatory, 1990). This construction resulted in a world-class astrophysical observatory: the Fred Lawrence Whipple Observatory and the unique Multiple Mirror Telescope that revolutionized optical astronomy.

It was at the dedication of the Whipple Observatory in 1968 that the first sign of changing public opinion manifested itself. Looking back in 1989, Representative Udall, who had played a key role in setting aside land in the Catalinas and Santa Ritas for astronomers, said, "Yes, I supported telescopes on Kitt Peak, I supported telescopes in the Santa Ritas. I was there when they dedicated Mt. Hopkins and then I said, 'That's enough.'" At the time, this warning appeared to be political rhetoric, but in light of later developments, astronomers should have taken it more seriously as an indication of shifting opinion.

In June 1984, astronomers at the university's Steward Observatory submitted a plan to the Forest Service to place up to 18 telescopes scattered over more than 1,200 ha (3,000 acres) of the Pinaleño Mountains, principally on Mt. Graham and Emerald Peak (see Chapters 1 and 2). The plan was ambitious, reflecting the conviction of the university that it was the "Wall Street of Astronomy": a claim not without foundation, since the university had an impressive array of installations in the area including the Steward Observatory, the university's Mirror Laboratory, the National Radio Astronomy Observatory, the National Optical Astronomy Observatory's facility at Kitt Peak, the Smithsonian's Whipple Observatory at Mt. Hopkins, the Mt. Bigelow Observatory, and the Flandrau Planetarium at the university.

After receiving the university's plans for astrophysical development in the Pinaleño Mountains, Forest Service administrators integrated the proposal into the ongoing analysis and development of the management plan for the Coronado National Forest. As the planning efforts of the Forest Service continued, however, significant public interest and the lack of site-specific data required a change in strategy. In June 1985, the supervisor of the Coronado National Forest decided to drop the astrophysical site analysis from the management plan and develop a separate and specific environmental analysis for what was then called the "Mt. Graham Astrophysical Area." This area (considered as part of the environmental impact statement), comprising 3,500 acres (1,400 ha) in the early stages of planning, was later modified to between 3,000 and 3,200 acres (1,200–1,300 ha).

In December 1985, the biologists with the Office of Arid Lands Studies at the University of Arizona provided the Forest Service with environmental data for the proposed astrophysical site, which were used extensively throughout the preparation of the environmental impact statement. In October 1986, Forest Service administrators released the draft environmental impact statement on the proposed Mt. Graham Astrophysical Area for public comment. Consistent with Forest Service practices under NEPA, the draft environmental impact statement considered six alternatives in addition to the Forest Service's preferred alternative. Each alternative furnished a different way to address issues, while providing for the use and protection of resources and compliance with legislative requirements. Forest Service administrators were required to use the information analyzed in the draft environmental impact statement to determine (1) the land allocation and management objectives desired on the 3,500-acre (1,400 ha) area; (2) the extent of development or nondevelopment of the area; and (3) the types of mitigation measures required under each alternative. The authors of the draft environmental impact statement noted, "While other mountains may be suitable sites for astronomical development, only the suitability of Mt. Graham for astronomical development is being considered. *Consequently, consideration of alternative locations is outside the scope of this analysis and decision*" (USFS 1986:3–99; emphasis added).

The draft environmental impact statement outlined nine areas of broad concern for analysis: plant and animal diversity, watershed management, recreation uses and opportunities, wilderness and special area designations, visual quality, cultural resources and Native American use, astrophysical values and benefits, socioeconomic impacts, and safety/protection. The preferred alternative called for the use of only seven acres for astrophysical development, thus allowing the university to develop no more than 5 of the 18 telescopes it had once envisaged: the 10 m Submillimeter Telescope (SMT), the Arizona/Ohio State Large Optical/Infrared telescope, the Texas 5 m telescope, and one small and one large optical/infrared telescope. The preferred alternative would not allow for the development of the National New Technology Telescope—originally the driving force behind the idea of an observatory on Mt. Graham—nor would it accommodate the Smithsonian's submillimeter array, due to the reduced acreage. The plan also allowed for the mixed use (especially for recreation) of the remainder of the Mt. Graham Astrophysical Area and called for the development of a recovery plan for the Mt. Graham red squirrel and formal consultation with the U.S. Fish and Wildlife Service on future land management. In light of subsequent developments, these last two stipulations can be considered the most significant recommendations of the draft environmental impact statement.

Concurrent with the listing of the Mt. Graham red squirrel as an endangered species in June 1987, the Forest Service initiated formal consultation with the Fish and Wildlife Service under Section 7 of the ESA, which requires federal management agencies to cooperate in protecting endangered species under the purview of the secretary of the interior. Researchers, resource managers, and environmentalists alike have paid considerable attention to the impact of the provisions of Section 7 of the ESA on the indirect management and conservation of wildlife.

One danger of the federal laws was that they may have been too effective. Representative John Dingell of Michigan, who along with his chief counsel Frank Potter drafted the 1973 Endangered Species Act, recognized both the potential desirability and the danger of Section 7 consultation. In 1978, Dingell and Potter wrote, "Without any of the guarded 'wherever practicable' or 'whether consistent with their other duties' language which characterizes earlier legislation of this type, this section requires that all federal agencies use their authorities in furtherance of the purposes of this Act. The purposes . . . are broad and they reach far. They may even reach too far; if they were to be used in the wrong case, to impose actions which would offend the wrong court, the clear mandate of the statute could well precipitate a repeal of this statutory language through overwhelming pressure upon Congress" (p. 305). Picking up on this theme, S. Dillon Ripley and Thomas Lovejoy commented, "In one sense, environmental groups are correctly using this legislation to block government projects which are detrimental to endangered species, but when the acts are used to combat all that is wrong about society's approach to environment, the legislation stands in danger of losing public support and being weakened" (1978:376). Indeed, as Michael Bean (1978:285) has pointed out, "the courts have become a focal point in the interpretation of wildlife laws," being used by environmentalists opposed to government projects (as shown below in the Mt. Graham lawsuits).

Nor is the issue merely protection of the environment versus development; another philosophical issue is Who shall govern? In the case of the Mt. Graham red squirrel, John Moag (1990), an attorney representing the university, wrote, "The ESA is perhaps the only federal statute that has removed elected officials and policy-makers from the decision-making responsibilities over the critical economic, social, and political issues affecting land-use rights and instead substitutes the judgment of a profession characterized by its fractiousness—namely, biologists. . . . But history has shown that biologists are neither necessarily objective nor independent, and the criteria must not be biological data alone." He further argued, supporting Dingell and Potter's prediction, that "the laudable and proud intentions of the ESA are not being met, they are being abused, and it is time to change the law."

The Biological Dimension of Observatory Negotiations

Regardless of protests about the excessive nature of the existing resource management legislation, the federal agencies were required to follow the established procedures of NEPA and the ESA. In late July 1987, they presented tentative "reasonable and prudent alternatives" (through the Forest Service) to the university. The university argued that the best site for its 11.3 m telescope and other optical instruments was not the Forest Service–preferred alternative of High Peak (Mt. Graham, *sensu strictu*), but Emerald Peak. However, later in the day university supporters of the observatory indicated their acceptance of High Peak. A few weeks later, the university, after considering the alternatives further, informed the Forest Service that the alternative under consideration (High Peak) "did not provide for or allow a viable cost-effective research facility" (Spear 1990). In other words, without more room for telescopes, astronomers felt that they would not be in a position to develop the MGIO.

In August 1987, the Fish and Wildlife Service provided a draft biological opinion on the red squirrel to the Forest Service, analyzing "reasonable and prudent alternatives" as required by the ESA, that is, analyzing whether any proposed alternative for the facility would likely jeopardize the continued existence of the species or result in the destruction or adverse modification of its critical habitat. The ESA and Fish and Wildlife Service regulations require the agency to "use the best scientific and commercial data available" in rendering a biological opinion (Spear 1990). The August 1987 evaluation noted that "given the [squirrel's] current severely endangered status, the loss of even a few acres could be critical to the survival and recovery of this species" (Spear 1987).

Regarding the Emerald Peak alternative that was the university's preference, the evaluation stated that no alternative could be developed that would remove the jeopardy to the species. In the section describing alternatives eliminated from consideration the draft biological opinion stated that regardless of the university's proposal to build a new shorter road to Emerald Peak and to close the existing road, building the facility there "would require clearing of spruce-fir habitats and would adversely affect middens both directly and indirectly through wind throw, solar drying and other effects. Destruction of habitat on Emerald Peak for siting an observatory would have greater detrimental impacts than the proposed siting on High Peak. Furthermore, those impacts on Emerald Peak could not be reduced below jeopardy with reasonable and prudent alternatives" (Spear 1987). This statement loomed large in future investigations surrounding the decision-making process for Mt. Graham.

In mid September 1987, the regional forest supervisor, Robert Tippeconic, concerned by the university's continuing negative reaction to the July 1987 alternatives, requested formal suspension of the Forest Service's consultation with the Fish and Wildlife Service under Section 7 to allow the university to present more information on the astronomers' new plans for a "minimum viable observatory." In early October 1987, the university presented plans for seven telescopes (four on Emerald Peak and three on High Peak) and necessary logistics and support facilities. The new proposal included no dormitories. Of the new telescopes proposed, five were large telescopes, one was a small telescope, and one was the proposed Smithsonian interferometer. The 11.8 m scope was to be placed on Emerald Peak. The proposed development affected 23 acres (9 ha) directly and 65–105 acres (26–42 ha) indirectly, as opposed to the June 1987 plan, which affected 9 acres (4 ha) directly and 71 acres (29 ha) indirectly.

On the basis of the university's revised plan, Tippeconic re-initiated

Section 7 consultation with the Fish and Wildlife Service in February 1988. That same month, the expanded biological assessment (Coronado National Forest 1988) of the proposed astrophysical development by the Forest Service also raised serious concerns about the fate of the red squirrel. The conclusions indicated that any change to the amount of habitat would be significant since both the size of the habitat and the population of squirrels were small and the "threshold of extinction" — i.e., the minimum number of squirrels needed to survive — was unknown.

In June 1988, the Fish and Wildlife Service presented the Forest Service Regional Office with two reasonable and prudent alternatives: (1) no telescopes at all, and (2) development of High Peak alone. Representatives of Steward Observatory (and thus the university) reiterated their desire that an 11.3 m telescope be placed on Emerald Peak and stated that otherwise the overall project would not be feasible. Fish and Wildlife Service officials, communicating through the Forest Service, stated that development of two peaks was impossible and furthermore that Emerald Peak was not contained in the university's plans of October 1987. The management agencies did reach agreement, however, in deciding to analyze a new road alternative to Emerald Peak.

On 14 July 1988, the Fish and Wildlife Service issued a jeopardy biological opinion (Spear 1988) with three reasonable and prudent alternatives. In addition to the two reasonable and prudent alternatives offered in June, a third allowed for three large telescopes on Emerald Peak, including the 11.3 m telescope sought by the university, along with a new access road. Facility development would affect 8.6 acres (3.5 ha) directly and 24–38 acres (10–15 ha) indirectly. The biological opinion stated that the university would reforest 60–80 acres (24–32 ha) of habitat previously cut by the Forest Service, pay the salary of a Forest Service biologist on the mountain, and fund and maintain a long-term monitoring program for the red squirrel to ensure its protection. An additional 17 acres (7 ha) would be restored over the long term by closing summer cabins at a site called Columbine and closing the Bible camp nearby. The opinion also stated that the project on Emerald Peak would make about 47 acres (19 ha) permanently unsuitable for squirrel habitat because opening the tree canopy would expose adjoining acres to heat and drying, thereby destroying food supplies needed by the squirrel to survive the winter.

Testifying before a congressional hearing examining the Mt. Graham decision in June 1990, Michael Spear, regional director of the

Southwest Region of the Fish and Wildlife Service, drew the congressional panel's attention to the second paragraph of the July 1988 biological opinion sent to the Forest Service: "This biological opinion is not a land use allocation analysis. The decision about the best use of the upper elevations of the Pinaleño Mountains involves issues beyond the [Endangered Species] Act and is ultimately your decision. In accordance with the Act, this opinion deals solely with the effect of the above proposed plans on endangered species." Further on, and with emphasis, Spear testified, "The opinion does not say there will be no adverse impacts. It says that, considering all of the factors which determine the eventual fate of the squirrel, the net changes brought about by the addition of the three scopes are not significant."

At the same hearing, James Duffus of the General Accounting Office, who led an investigation (at congressional direction) on the biological opinion, testified that Spear told the investigators he had considered several other factors in reaching his decision, including "(1) the need to make an expeditious decision; (2) the University's vigorous insistence on the Emerald Peak area as the only viable site for the facility; (3) his belief that the University would win in a court of law; and (4) his perception that one of the telescopes represents a world-class scientific development." Duffus also testified that Spear had stated, "The challenge of the Endangered Species Act is to devise compromises that accommodate both needed projects and endangered species."

The General Accounting Office had two primary concerns about how the biological opinion had been prepared. The first concern was that Fish and Wildlife Service biologists had rejected the alternative allowing three telescopes on Emerald Peak in preparing the draft biological opinion. In January 1990, these biologists' depositions before the General Accounting Office were leaked to the press, revealing allegations that Spear had directed the biologists to write alternative three into the final biological opinion to permit construction of the observatory. This testimony provoked outrage in the environmental community and led to an aggressive lobbying campaign on Capitol Hill in an attempt to overturn the provisions of the Arizona-Idaho Conservation Act, which had authorized the observatory (see below), and stop further construction.

The second concern of the General Accounting Office was that Spear had considered "nonbiological information in reaching an opinion that could jeopardize a species existence" (Duffus 1990). Duffus testified, "We believe now, as we did then, that biological decisions should be based on biological information. Weighing the risk of a species' extinction with the benefits of a project is a policy decision and should be left to a high-level Endangered Species Committee whose members are Secretaries of Interior, the Army, and Agriculture; the Chairman of the Council of Economic Advisers; the Administrators of the Environmental Protection Agency and the National Oceanic and Atmospheric Administration; and a representative from the affected state to whom the act assigns responsibility for granting exemptions to the Act's protective provisions." This so-called God Committee (or "God Squad") was convened by the Bush administration to consider a BLM petition to log federal land in Oregon and then convened by the Clinton administration to adjudicate the northern spotted owl controversy in the Pacific Northwest.

Ensuing Political and Legal Actions

However, the 1990 investigation and congressional hearings were neither the first nor the most controversial response to the jeopardy biological opinion. Information from the Fish and Wildlife Service and the Forest Service indicated that additional studies, appeals, and other requests for consultation could take many more months or even several years before issuance of a final decision to proceed, or not, with construction. The university, already concerned that the normal Section 7 consultation period of 90 days had taken nearly two years, feared that the delay in issuing a final decision in the form of a final environmental impact statement would jeopardize the economics of the partnerships entered into and could result in the international and national consortia deciding to take their case directly to the Congress of the United States (personal communication from Margaret A. McGonagil of the University of Arizona, 1990).

From the beginning, the university had diligently kept the Arizona congressional delegation informed of its plans. Using its Government Relations Office, representatives of the university met frequently both in Arizona and in Washington with the state's representatives and senators. The most actively interested legislator was Senator Dennis DeConcini (D-Ariz.), who in 1987 recognized that solutions to the conflicts over NEPA and the ESA might require congressional relief.

In early 1988, the university hired John Moag, a lobbyist from the prestigious Washington, D.C., law firm of Patton, Boggs, and Blow, to represent the university on a number of issues before Congress, one of these being Mt. Graham. Moag realized that the key member of the Arizona delegation needing to be won over to the university's position was Morris Udall, the chairman of the powerful House Interior Committee. Without Udall's support any plans for Mt. Graham on the university's part were doomed. Consultations began in earnest with Udall's key aide, Mark Trautwein. The university sought Trautwein's assistance in persuading the Fish and Wildlife Service to side with the university. Trautwein, in an interview with Charles Bowden in *City Magazine* of Tucson, recalled a heated exchange in a meeting with university officials, and he made it clear that he would recommend no assistance on this matter to Udall. The rest of the Arizona delegation was prepared to assist, but in deference to Udall withheld specific legislative relief at the time.

Over the Fourth of July weekend in 1988, university representatives met again with the Arizona delegation in anticipation of the final biological opinion of the Fish and Wildlife Service and argued for specific legislative relief. Senator DeConcini and his staff began the process of drafting legislative amendments to satisfy the university's concerns. In the process, Moag and university officials met with the House Merchant Marine Committee (concerned with oversight of the ESA) to discuss their views. They also met with members of the Senate and the Environment and Public Works Committee. The key, however, remained Udall.

Finally, on September 28, Henry Koffler, president of the university, met privately with Representative Udall; following that meeting, Udall indicated to his staff that he would be "an open door not a brick wall" (Bowden 1989). Although Udall argued that this was a scientific matter and not a political one, the university argued that in fact the situation had become political and that legislative relief was the only sensible option.

National administrators of the Fish and Wildlife Service and Forest Service were consulted on the proposed legislation, and after being assured that the university would abide by the conditions set under the third alternative of the biological opinion, they agreed not to oppose the legislation. Further, the administrators of these federal agencies made clear to various congressional committees that although construction of three telescopes would be permitted with NEPA and ESA waivers, the additional four proposed for Emerald Peak later on would be subject to the normal regulatory processes under both NEPA and the ESA. Moreover, any change in the red squirrel population during the construction period of the first three telescopes would trigger a reconsultation process between the Fish and Wildlife Service and the Forest Service under Section 7 of the ESA.

With these conditions agreed upon, Senator DeConcini attached an amendment to pending legislation involving the transfer of significant environmental properties in Arizona and Idaho. The amendment permitted construction of three telescopes on Emerald Peak and allowed for possibly four additional telescopes following appropriate environmental review. On the floor of the House of Representatives, Representative Udall defended the package, although he said "it is hard, to think of any recent environmental issue in Arizona that has stirred more genuine emotion and heated controversy than this one" (*Congressional Record* 1988). The legislation passed both houses of Congress overwhelmingly and became the Arizona-Idaho Conservation Act of 1988. A few weeks later the Forest Service issued the final environmental impact statement, thus fulfilling the preconstruction requirements of NEPA (USFS 1988).

In April 1989, the Forest Service and the university signed the special use permit and adopted an MGIO management plan, which had been developed in February 1989 and had been subjected to public comment during a relatively quiet comment period. On 27 July 1989, a coalition of environmental organizations filed suit in U.S. district court in Arizona against the secretary of agriculture alleging violations of the ESA and the Arizona-Idaho Conservation Act as detailed below. The case, entitled *Mt. Graham Red Squirrel, et al. v. Yeutter, et al.,* was assigned to Judge Alfredo C. Marquez. In autumn 1989, the district court denied the motion of the plaintiffs to grant a preliminary injunction to bar construction of the proposed access road to the observatory site on Emerald Peak. However, in March 1990, in response to another motion by the plaintiffs, the district court granted a 120-day injunction halting construction of the entire project. Then, on 15 May 1990, the Ninth Circuit Court of Appeals stayed the 120-day injunction and said in its stay order, "It appears that plaintiffs' claims under the Endangered Species Act and the Arizona-Idaho Conservation Act do not raise a serious question." The district court on 4 June 1990 granted, in an order signed by Judge Marquez, the U.S. government's motion for summary judgment on all but two of the plaintiffs' claims, as detailed below.

In challenging the Forest Service's issuance of a special permit to the university, the plaintiffs raised a number of legal issues. Among these were the failure to designate critical habitat in a timely fashion, enlargement of the project area beyond that originally considered, the Forest Service's management practices, the taking of squirrels in violation of the Arizona-Idaho Conservation Act, and failure to re-institute consultation under Section 7 of the ESA.

The district court, in considering the plaintiffs' motions, entered a summary judgment on their claim that the government had failed to designate critical habitat under Section 4 of the ESA: the court pointed out that on 5 January 1990 the Fish and Wildlife Service had made a final designation of critical habitat that was identical to the boundaries proposed in 1986 and upon which the biological opinion of 1988 had been based. In the court's view, this action made moot the plaintiff's claim of failure to designate critical habitat. Similarly, the district court also entered a summary judgment against the plaintiffs on their claim that the Forest Service's forest and wildlife management practices and the enlargement of the project beyond the 8.6 acres for telescopes had resulted in the taking of squirrels in violation of Section 9 of the ESA. The court's view was that the plaintiffs had not been successful in identifying any particular management practices that had resulted in identifiable takings, nor had these takings exceeded the limit of six per year established under the third alternative in the biological opinion. As far as the enlargement issue was concerned, although university astronomers had at one point asked that the area be enlarged, they later withdrew this proposal. Thus, the court found both these complaints groundless.

The court also granted summary judgment against the plaintiffs on their claim that the Forest Service had failed to take effective measures to conserve and ensure a viable population of the Mt. Graham red squirrel in violation of the ESA and the regulations promulgated under the National Forest Management Act. In implementing the terms and conditions of the third alternative in the 1988 biological opinion, the Forest Service had banned logging, Christmas tree cutting and fuelwood gathering, and grazing in the habitat of the red squirrel. In addition, the Forest Service was closing roads and reforesting areas in the Pinaleños. The Forest Service also required, as part of the special use permit for the observatory, 20 specific features detailed in the third alternative of the biological opinion that protected squirrel habitat.

The plaintiffs also claimed that the Fish and Wildlife Service and the Forest Service should have re-initiated consultation under Section 7 of the ESA due to enlargement of the project, declining squirrel populations, and failure of the Fish and Wildlife Service to designate critical habitat. Two of these claims had already been resolved; as to the population decline, Section 7 allows for re-initiation of consultation if new information reveals effects of the permitted action that may affect the listed species or critical habitat in a manner or to an extent not considered in a biological opinion. The court considered the previous evidence, which had indicated that the squirrel populations fluctuate widely, and felt that the population information cited by the plaintiffs did not constitute new information. The district court and the court of appeals also considered the testimony of two biologists with the Fish and Wildlife Service who alleged that the third reasonable and prudent alternative in the 1988 biological opinion was predetermined (see Chapter 2). After considering the sworn depositions of the two biologists, the courts ruled that their charges did not represent new information, and thus reinstitution of consultation was not necessary.

While the legal battles proceeded, in August 1989 the 30-day baseline preconstruction monitoring program on the red squirrel had begun. Over the year, two red squirrels died as a result of stress during handling, which led opponents of the observatory to take even more aggressive action. In October 1989, the university began construction of the new access road amidst opposition demonstrations on the mountain. The total area for the road would be 5.6 acres (2.3 ha). Half the road was classified as poor habitat and was outside the squirrel refugium. The one mile of road within the refugium was aligned to avoid all middens.

In February 1990, Tippeconic notified the Fish and Wildlife Service, under the provisions of the ESA, that since critical habitat for the red squirrel had been designated, he could find no authority in the Arizona-Idaho Conservation Act requiring further consultation over the first three telescopes. Later in the same month, Forest Service administrators indicated that, in their view, re-consultation should involve only the fourth through seventh telescopes. In March 1990, the regional director of the Fish and Wildlife Service responded expressing disagreement on the Forest Service's interpretation of the Arizona-Idaho Conservation Act in regard to re-consultation. Representatives of the two agencies agreed that if they could not resolve this issue, they would ask the U.S. Justice Department to render an opinion. Concerned about further delays in construction, the university vigorously backed the Forest Service's position.

In May 1990, administrators of the Phoenix office of the Fish and Wildlife Service met with representatives of the Forest Service to discuss the construction plan for the first two telescopes on Emerald Peak and reviewed issues relating to residence facilities and determination of acreage used for completion of the new access road. In the same month, the chairman of the House Merchant Marine Committee, Walter Jones, and his subcommittee chairman on Fisheries and Wildlife Conservation, Gerry Studds—who exercise oversight for the ESA—wrote to Max Petersen, the chief of the Forest Service, to express their opinion that Title VI of the Arizona-Idaho Conservation Act did not merit waiver of the Forest Service's obligation to re-initiate consultation with the Fish and Wildlife Service in accordance with Section 7. Other members of Congress sent similar letters to Petersen. Senators Dennis DeConcini and John McCain of Arizona, proponents of the project, asked the General Accounting Office (the legislative watchdog) to undertake an investigation of the 1988 biological opinion of the Fish and Wildlife Service. Bowing to mounting pressure from several constituencies, the House Interior Committee (then chaired by Morris Udall of Arizona) and the House Merchant Marine Committee decided to hold a joint hearing on the 1988 biological opinion.

The hearing was held on 26 June 1990 before an overflowing audience with Bruce Vento, chair of the House Subcommittee on National Parks and Public Lands, and Gerry Studds presiding. Also present at various times were representatives of the Arizona House delegation, including Representative Udall (who was by then very ill) and Senator McCain. The hearing lasted almost a full day, receiving testimony from representatives of the General Accounting Office, the Department of Justice, the Fish and Wildlife Service, prominent environmental groups, and the university. At times the hearing took on the trappings of a legal proceeding, with congressional supporters of the observatory project serving as "prosecutors" to those witnesses not predisposed to their viewpoint. At other times, representatives of the university and the Fish and Wildlife Service were sharply questioned. Reams of testimony prepared in advance of the hearing were available, although these were largely ignored in the process. Throughout the entire day, scientific findings and opinions were expressed but, as an illustration of how minimally science informed the process, members of the House asked very few follow-up questions concerning data. At this point, the matter had come down to arcane legal interpretations of both NEPA and the ESA. The Justice Department, in outlining its position, relied heavily on the decisions already rendered in the circuit court affirming the Forest Service procedures.

Several years behind in their efforts to inform Congress of their views, the environmental groups attempted to outline their position from the beginning of the project, with a dizzying array of statistics and data which, too, were largely ignored. The university presented its position both on political and scientific grounds but reminded the Congress that this matter was now in the hands of the courts. The hearing ended with a whimper, with no sense of resolution, as most members of the House moved off to other hearings. However, following the hearing, four Arizona congressional representatives issued a press release asking that the university review biological data for the red squirrel and suspend further construction until such information was properly brought up to date.

In a statement released on 28 June 1990, the president of the university, Henry Koffler, offered the view that "the hearing was objective and substantively addressed the process leading up to the enactment of the Arizona-Idaho Conservation Act. Testimony from the U.S. Justice Department, the U.S. Fish and Wildlife Service and the U.S. Forest Service indicated that there was no legal or biological basis at this time for a reconsultation of the project." However, he added, "While the University has the legal right to proceed with the project, it will do so only in deference to the wishes of the members of the Arizona Congressional delegation." Later in the same statement, Koffler expressed concern about the ramifications of further delay in terms of the economic viability of the project and its implication for keeping together the coalition of organizations interested in developing the observatory.

Following further consultation with Congress and federal agencies (and backed by federal court rulings), in the late summer of 1990 the university resumed construction on the road. Members of environmental groups, primarily those from the region, continued a letter-writing campaign and conducted other protests. Several potential partners removed themselves from participation in the observatory development, but these decisions were based largely on either scientific grounds, as in the case of the Smithsonian Institution's decision to place its submillimeter array on Mauna Kea, or on economic grounds, as in the case of Ohio State University, which withdrew due to lack of state funding.

Lessons

Many lessons can be drawn from the Mt. Graham affair. First among these is that "biological truths" are hard to determine absolutely. The time frames established under NEPA and the ESA do not coincide with time frames for biological investigations necessary to assess adequately the jeopardy to a species and its habitat. Thus, biologists must make a judgment call at a particular moment in time and hope that their capacities for prediction, which are admittedly imprecise, will prove to be correct (cf. Chapter 2).

Second, agencies of government that manage public lands must deal with a crazy-quilt of conflicting laws and regulations. The Fish and Wildlife approach to preservation clearly conflicts with the multiple-use philosophy of the Forest Service. Federal resource managers and Congress need to overhaul the existing set of laws as well as agency-generated regulations and policies (largely developed ad hoc in the late nineteenth and early twentieth centuries) and consolidate administration in one agency with a set of laws that allows for proper management and economic growth, while reflecting modern-day biological standards and adequate conservation.

Third, the present use of environmental impact statements (EISs) allows the proponents of development to marshal considerable human and monetary resources to make their case. The agencies who administer NEPA are faced with funding constraints and timelines that do not allow an adequate assessment of all information. Consequently, the process of producing an environmental impact statement has become routine and largely reliant on information provided by the proponents rather than the agencies.

Fourth, land managers should make their decisions on the basis of statutory criteria; if these decisions are found to be unpopular, it becomes incumbent upon the politicians (and not the land managers) to suggest appropriate adjustments. When agencies attempt to enter the political arena, their credibility as stewards of the public trust is severely undermined.

Fifth, environmental organizations must reform the manner in which they raise funds. The attractiveness of generating publicity for various projects (some based on solid data and others not) to tie into a fundraising campaign brings into question the credibility of their organizations. A number of regional and national organizations took advantage of the Mt. Graham situation for fundraising without participating in the actual policy debate.

Sixth, biologists need to reconcile biological imperatives (i.e., conservation) with other socially desirable issues. Biologists do not exist in a vacuum and must express their views within a larger context and in manner intelligible to the general public. When biologists feel strongly about an issue, they must overcome their traditional caution and make their views known earlier. They must also let decision-makers know that their judgments are often made on the basis of the "best available data" and that choices sometimes have to be made on less than complete information. Biologists must also recognize that they have important contributions to make to decisions on land management and development.

Seventh, astronomers need to recognize that the future of ground-based astronomy depends upon their ability to articulate their case earlier and better to a wide range of constituencies. In developing plans for new sites, a plan that includes biological, cultural, and physical characteristics should become a requirement, not an option. Since most desirable sites for new observatories are likely to be environmentally sensitive, a greater appreciation of the importance of biological considerations must become part of their consciousness.

Eighth, universities or other nonprofit institutions involved in the NEPA and ESA processes must give greater consideration to getting their plans out early and making certain that a maximum number of groups become involved in the resulting public dialogue. Consistency of approach and the commitment to an "open process" must be absolute. Such institutions should not think that their special niche in society relieves them of legal and moral obligations. New techniques in "environmental mediation" might also be considered before issues reach the point of local or national controversy.

Ninth, politicians and the courts are, in the end, the arbiters and ultimate stewards of our public lands. They must recognize that abrogating laws through legislative devices undermines their credibility. With the public, the "ultimate property owners" in our democratic system, they must approach these issues with an open mind and let the process run its full course. Only then should consideration of means of intervention be sought and only under extraordinary circumstances.

As in most cases involving complex and sensitive issues, history will be the final judge of this affair. Mistakes were made on all sides. The application of our environmental laws and their intent are subject to multiple interpretations. The saga of Mt. Graham is perhaps one of the best illustrations of modern democracy at work—for better or for worse. The environment has become a national touchstone, and its future health requires an informed constituency. Perhaps the greatest lesson to be learned is that all citizens need to be better informed of environmental laws and regulatory processes to participate in national debates over the protection and use of natural resources — in the many debates yet to come.

References

Bean, M. J. 1977. *The Evolution of National Wildlife Law.* Superintendent of Documents, U.S. Government Printing Office, Washington, D.C.

Bowden, C. 1989. How the University Knocked Off Mt. Graham. *City Magazine.* January 1989, p. 29.

Brokaw, H. P., ed. 1978. *Wildlife and America: Contributions to an Understanding of American Wildlife and Its Conservation.* Superintendent of Documents, U.S. Government Printing Office, Washington, D.C.

Congressional Record. 1 February 1984, S-685. 98th Cong., 2d sess.

Congressional Record. 20 October 1988, H-10546. 100th Cong., 2d sess.

Coronado National Forest. 1988. *Mount Graham Red Squirrel: An Expanded Biological Assessment.* Coronado National Forest, Tucson, Ariz.

Dingell, J. D., and F. M. Potter. 1978. Federal Initiatives in Wildlife Management. In *Wildlife and America: Contributions to an Understanding of American Wildlife and its Conservation,* ed. H. P. Brokaw. Superintendent of Documents, U.S. Government Printing Office, Washington, D.C.

Duffus, J., III. 1990. General Accounting Office. Statement before the House Merchant Marine Subcommittee on Fisheries and Wildlife Conservation and the Environment and the House Interior Subcommittee on National Parks and Public Lands.

Gippert, M. J., and V. L. DeWitte. 1989. Forest Plan Implementation: Gateway to Compliance with the Forest Management Act, the National Environment Policy Act, and Other Federal Environmental Laws. (Unpublished.)

Koffler, H. 1990. Statement Regarding the Mt. Graham International Observatory, University of Arizona, Tucson. 28 June.

Moag, J. 1990. It Is Time to Amend the Endangered Species Act. *Washington Legal Foundation* 5(40). 19 October.

Mt. Graham Red Squirrel, et al. v. Clayton Yeutter. 1990. U.S. Federal District Court for the District of Arizona, Tucson, Arizona. Case #CIV 89-410 TUC ACM. Later changed to *Mt. Graham Red Squirrel, et al. v. Edward R. Madigan, et al., and State Board of Regents, University of Arizona.*

Ripley, S. D., and T. E. Lovejoy. 1978. Threatened and Endangered Species. In *Wildlife and America: Contributions to an Understanding of American Wildlife and its Conservation,* ed. H. P. Brokaw. Superintendent of Documents, U.S. Government Printing Office, Washington, D.C.

Spear, M. J. 1987. *Draft Biological Opinion.* U.S. Department of the Interior, Fish and Wildlife Service, Albuquerque, N.M.

Spear, M. J. 1988. *Biological Opinion.* U.S. Department of the Interior, Fish and Wildlife Service, Albuquerque, N.M.

Spear, M. J. 1990. Statement before the House Merchant Marine Subcommittee on Fisheries and Wildlife Conservation and the Environment and the House Interior Subcommittee on National Parks and Public Lands.

USFS. 1986. *Draft Environmental Impact Statement, Proposed Mt. Graham Astrophysical Area, Pinaleño Mountains, Coronado National Forest.* U.S. Department of Agriculture, Forest Service, Southwestern Region, Tucson, Ariz.

USFS. 1988. *Final Environmental Impact Statement, Proposed Mt. Graham Astrophysical Area, Pinaleño Mountains, Coronado National Forest.* EIS #03-05-88-1. U.S. Department of Agriculture, Forest Service, Southwestern Region, Tucson, Ariz.

2

Modern Astronomy

The Purpose and Promise of the Mt. Graham International Observatory

We find ourselves in a bewildering world. We want to make sense of what we see around us and to ask: What is the nature of the universe? What is our place in it and where did we come from? Why is it the way it is? . . . But ever since the dawn of civilization, people have not been content to see events as unconnected and inexplicable. They have craved an understanding of the underlying order in the world.

S. W. HAWKING, *A Brief History of Time*

The chapters in Part 2 survey the field of astronomy as background for understanding the scientific goals of the Mt. Graham International Observatory and explain the special aspects and advances in astronomy that it is designed to make possible. The first paper, by Martin Harwit, provides a historical perspective and discusses the future need for ground-based astronomy on mountaintops. In the second paper Peter Strittmatter describes the main features of the "new astronomy," the advanced technology and instruments involved, the scientific goals for future astronomical research, and the process leading to selection of the Pinaleño Mountains as an astrophysical site. He also mentions areas where he believes astronomers and biologists have much in common.

CHAPTER FOUR

Astronomy Without and With Access to Mountaintops

Martin Harwit

In 1845, William Parsons, third earl of Rosse, completed the construction of a telescope six feet in diameter at Birr Castle, Parsonstown, in Ireland. It was so much larger than any other telescope that had ever been constructed or would be constructed during the remainder of the century that it came to be known as the leviathan of Parsonstown. No telescope surpassed it in size until the early decades of the twentieth century. Within two months of its erection the telescope showed up a curious spiral structure in certain nebulosities. Today, we know that these are spiral galaxies, similar to our own Milky Way. But that interpretation did not become firmly established until around 1925.

Despite its enormous size, and the discovery of spiral nebulae, Parsons's instrument was a disappointment. The elevation of Birr Castle turned out to be far too low to permit full use of the power that such a telescope could have offered. Looking back at the turn of the century, Agnes Clerke, arguably the best historian of astronomy of her times, reported:

> No sooner had the Parsonstown telescope been built than it became obvious that the limit of profitable augmentation of size had, under climatic conditions at all nearly resembling those prevailing there, been reached, if not surpassed; and Lord Rosse himself was foremost to discern the need of pausing to look round the world for a clearer and stiller air than was to be found within the bounds of the United Kingdom. . . . Professor Piazzi Smyth's experiences during a trip to the Peak of Tenerife in 1856 in search of astronomical opportunities gave countenance to the most sanguine hopes of deliverance, at suitable elevated stations, from some of the oppressive conditions of

> low-level star-gazing. . . . Now, at last, . . . mountain observatories are not only an admitted necessity but an accomplished fact. (1908: 434)

With this last sentence, Clerke was acknowledging that astronomers at the end of the century were systematically journeying to distant mountain sites to take advantage of the far higher clarity of the skies offered at the higher altitudes. S. I. Bailey of Harvard, for example, traveled to rugged Arequipa in Peru and was rewarded for his pains with the discovery of 85 rapidly pulsating stars among the hundreds of thousands of stars that populate the globular star cluster Messier 5. Other astronomers were building entire observatories on the tops of mountains. One of the most productive of these was the Lick Observatory on Mt. Hamilton in California. Such sites soon proved indispensable if astronomers were to make the best use of the large telescopes required for probing the universe out to extreme distances and back to the earliest times.

Mountains permitted rising above low-level clouds and above a large percentage of the turbulence that causes the images of stars to dance around, producing the scintillations astronomers call "seeing." Seeing blurs photographic images: stars that should show up as crisp, pointlike dots on a photographic plate instead spread into blobs that meld and smudge the appearance of galaxies and clusters. Detail is lost, and insight is thwarted.

Mountaintop observations provide for clearer radio and infrared as well as visual detection. At wavelengths around 20 micrometers—wavelengths thirty times longer than the eye can readily detect—astronomical observations are impossible unless astronomers use telescopes and infrared (thermal) sensing equipment stationed at least at the elevations of the highest, driest mountaintop sites. Atmospheric water vapor absorbs so much radiation from astronomical sources at these wavelengths that even normal mountaintop sites lie below too much moist air to permit such observations. Only a few of the very best sites will serve. Observations in the submillimeter domain, where wavelengths lie partway between infrared and radio wavelengths, also require such sites.

If observations carried out at different wavelengths all gave equivalent information about astronomical sources, none of this would matter. We would simply carry out astronomical observations at the most convenient wavelengths at our disposal, most likely observations at visual or radio wavelengths where the atmosphere offers the fewest limitations. But whereas visible light gives us information about ordinary stars, it teaches us nothing about sources such as the center of the Milky Way, because dense clouds of dust floating in the vast interstellar spaces between the center and our own location in the galaxy totally absorb all light. Radio and submillimeter radiation, by contrast, pass almost unhindered; and infrared observations, in turn, permit us to study the nature of the obscuring dust grains, their chemical properties and molecular structure. By the same token, infrared and submillimeter observations are not particularly suitable for studying ordinary stars that the human eye has no difficulty at all in seeing. Each wavelength at which we develop observing capabilities appears to teach us something quite new about the universe. Each needs to be pursued to obtain a complete picture.

Unfortunately, even the tallest mountains do not permit us to detect every kind of radiation. X-rays and most ultraviolet radiation, for example, are so strongly absorbed by atmospheric constituents that one has to go to far greater altitudes to detect these kinds of radiation from celestial sources. And so we build X-ray and ultraviolet orbiting observatories in space, where no atmosphere exists.

But building telescopes and instrumentation for space is immensely expensive. The mirror of the Hubble space telescope is not much larger than the mirror the earl of Rosse constructed. And although Rosse was by no means a poor man, he never could have afforded to spend the two billion dollars that the Hubble space telescope has cost.

Sending a telescope into space is not an affordable option for any telescope that can be operated on mountaintops. The space launch alone costs in excess of a hundred million dollars for large, telescope-sized payloads, and the overall cost of missions conducted in space tends to run a hundred to a thousand times higher than the expense of corresponding missions carried out from the ground. Were astronomers to give up the option of carrying out astronomical observations from the tops of mountains, the growth in our understanding of the universe would be impeded immeasurably, as costs would rise to unacceptably high levels. Acceptable alternatives to mountaintop astronomical observations are currently nowhere in sight.

References

Clerke, A. M. 1908. *A Popular History of Astronomy During the Nineteenth Century.* 4th ed. Adam & Charles Black, London.

CHAPTER FIVE

Astronomy and the Pinaleño Mountains

Peter A. Strittmatter

Until rather recently, astronomy was primarily concerned with exploration and documentation of the visible universe as revealed by observations of the night sky. Throughout recorded history, astronomers have tried to map—in three-dimensional space—the positions of cosmic phenomena, to classify them according to their appearance, and to catalogue their measurable properties. This form of primary exploration and cataloguing is not unlike that of botanists and zoologists during approximately the same time period. For both sciences, this cataloguing activity continues today but within an overall landscape that has become fairly well defined.

Astronomers also hope to understand, in terms of the laws of physics, chemistry, and so forth, the nature of the different observed phenomena: planets, stars, galaxies. Because astronomy is an observational rather than an experimental science, such understanding comes largely from the interplay between observation and theoretical modeling. Information flows both ways. For example, measurements of the motion of the planets by Tycho Brahe and others at the end of the sixteenth century led to Sir Isaac Newton's formulation of the laws of gravity about a century later. These laws, in turn, led to the ability not only to predict the motions of planets, but also to understand a vast number of other phenomena—including falling apples, the motions of projectiles, and the structures of stars and galaxies. Similarly, spectroscopic observations of stars have led to advances in atomic and nuclear physics just as observations in atomic and nuclear physics have led to advances in astronomy, in particular our understanding of the chemical composition of the universe. Today a similar interplay is taking place between observations of the early universe and elementary particle physics.

Beyond the goals of cataloguing and arriving at an understanding of natural laws, astronomers hope to integrate this knowledge into a "unified" picture of the origin and evolution of the universe, including the origin of the sun and planets and of the chemical elements. These are essential ingredients to understanding the origin of life on Earth—surely a question of great interest to biologists.

I suspect that astronomers and biologists have much in common in that they both observe very complex phenomena, which they classify according to appearance and try to understand in terms of simplified models. In a sense, they are at different extremes of a spectrum. Astronomical phenomena are generally so distant and so massive that experimentation is impossible; the biologist's observations, by contrast, are so near that biologists have difficulty in avoiding altering the experiment by human presence.

The Tools of the Astronomer

Astronomers acquire almost all their information by collecting electromagnetic radiation. The only exceptions are space probes sent to analyze objects in the solar system, the detection of cosmic rays (highly energetic particles that reach the earth from outer space), and the detection of neutrinos from the sun and certain special events like the 1987 supernova named 1987a. Electromagnetic radiation must be collected with some kind of imaging system (telescope, antenna, etc.). In general, the larger the collector, the greater the sensitivity and the better the limiting sharpness of the images it can obtain. Also, for the same size telescope, the limiting image sharpness increases toward shorter wavelengths.

Until around 1945, astronomers obtained almost all astronomical information from observations in optical light—a window in the electromagnetic spectrum covering roughly a factor of two in wavelength centered on green light (0.5 μ; 2.4eV: see figure 5.1). Since then, astronomy has undergone a complete revolution, as technology has increased the accessible wavelength range by more than a trillion—from radio waves at the lowest wavelengths to gamma rays at the highest, including infrared, ultraviolet, and X-ray radiation. Each wavelength range has revealed entirely new phenomena; for example, quasars and pulsars were discovered at radio wavelengths, stars in the process of forming have been revealed in millimeter and infrared light, and X-ray astronomy has led to the detection of black holes. Other wavelength ranges, such as the submillimeter region, have remained essentially

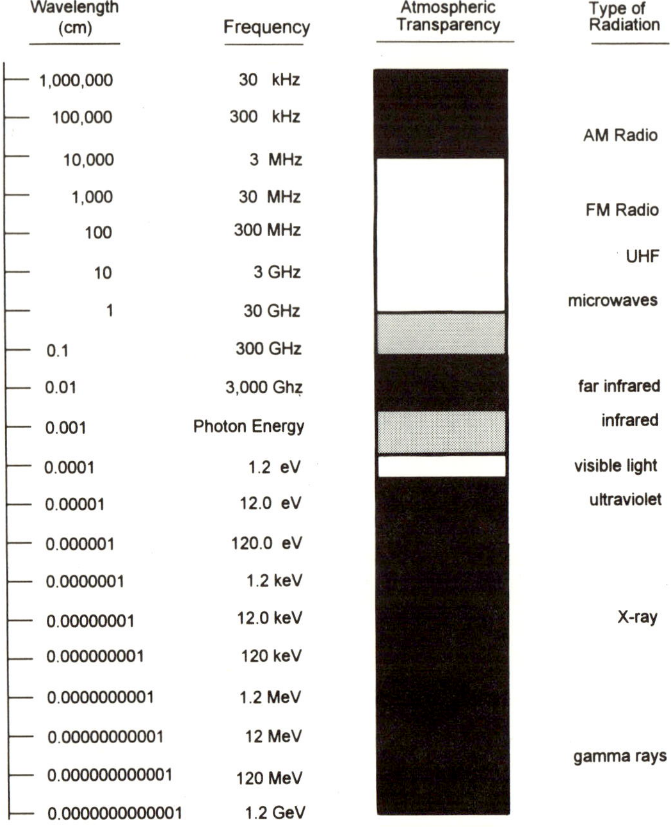

Figure 5.1 The various regions of the electromagnetic spectrum are shown in order of decreasing wavelength. Also shown is the atmosphere transparency: black regions are opaque, white regions are transparent, and gray regions are transparent provided the water vapor column density in the line of sight is low.

unexplored largely for technological reasons; however, they too offer a rich return in new knowledge.

Not all wavelengths are accessible to ground-based observations because of the opacity of the Earth's atmosphere. The atmosphere absorbs all wavelengths shorter than that of visible light—an essential condition for the survival of living organisms. The cutoff at roughly 300 nm is due to ozone. The atmosphere also absorbs wavelengths in various infrared bands—often due to water vapor, carbon dioxide, and other kinds of molecules. Within these wavelength regions, astronomers can make observations only from above the Earth's atmosphere—

from Earth orbit or from the Moon. Ideally, all astronomical observations would be made from such a location. However, the cost of a ground-based observatory is about a thousand times less than a similar space facility, even without counting launch costs. Therefore, whenever it is possible, astronomers make their observations from the ground (see Chapter 4).

The wavelengths accessible from the ground are the radio region, for which even rain is not a serious impediment; the millimeter and visible regions, for which clear skies are essential; and the submillimeter and infrared regions, for which, in addition to clear skies, low water vapor column densities in the line of sight are required. As a result, Arizona, Hawaii, and northern Chile have become favored sites for astronomy because they have clear and dry skies and relatively high mountains that rise above most of the dust, aerosols, and water vapor.

To increase observing sensitivity or power, the astronomer may either develop more sensitive or smarter focal plane instrumentation or increase telescope aperture. For submillimeter astronomy, instrumentation is still in its infancy; only within the last few years has it become possible to build telescopes and detectors suitable for this range. Optical and infrared telescope design, by contrast, has remained essentially stationary for over 50 years, since the 5 m Hale telescope was conceived. In the meantime, detector sensitivity has been increased by more than a factor of 50. Today's light sensors are very close to achieving 100% efficiency so that further gains along these lines are no longer possible. The only way to achieve higher sensitivity for these wavelengths is by constructing larger telescopes.

Until 1980, astronomers thought this too was impossible or, at least, ridiculously expensive. At that time, the commissioning of the six-mirror multiple mirror telescope, a joint venture of the Smithsonian Institution and the University of Arizona, demonstrated one way in which much larger telescopes could be built. The success of this new type of telescope has resulted in the present worldwide efforts of optical and infrared astronomers to construct telescopes in the 8–12 m class, increasing sensitivity by a factor of 4 to 100 over the present generation of "large" telescopes, the precise gain depending on the application. Approximately ten such instruments either are under consideration or are being constructed—despite the advent of the Hubble space telescope, which is actually very small in comparison but is uniquely capable of making observations in the ultraviolet region beyond the ozone cutoff in atmospheric transmission.

The new generation of large telescopes will permit exploration of conditions in the early universe. These new telescopes will allow study of the processes of galaxy and chemical element formation, looking back over more than 90% of the age of the universe. The larger telescopes will also yield unprecedented spatial resolution in the infrared region and hence new tools for the study of star and planet formation — as well as for the search for planets around other stars.

Site Requirements for the Next Generation of Telescopes

To be successful, the new large telescopes need to be placed on appropriate sites. The primary requirements for these telescopes are clear skies (free of clouds and dust), low water vapor column density, absence of artificial light sources, and low wind speeds. These conditions are best satisfied on mountains above 3,000 m in elevation in desert regions such as those offered by Arizona or Chile. They can also be found at very high altitude island sites such as Hawaii. From purely atmospheric considerations, the higher the site the better at any latitude. Indeed, the Moon would be the most favored site—high by any standard! This elevation issue, however, raises another set of criteria, namely, accessibility, cost of construction and operation, availability of technical resources, sociopolitical stability, and sky coverage (which increases toward lower latitudes).

Although it would be easy to discount these requirements, they can be crucial features in determining whether a project is viable. Obviously, if all the resources must be put into logistics, the astronomical goals will not be achieved, no matter how wonderful the site. For this reason, none of the large telescope projects will be located on the Moon. On a more local scale, the higher the altitude of the site, the harder it is to operate. The reasons are various. Above approximately 3,000 m, there is a very rapid fall off in human mental performance in the absence of long periods of acclimatization. At any latitude, the problems of snowfall increase with altitude. Thus the actual selection of a site is inevitably a compromise, with the final choice often depending critically on the specifics of the project.

In the early 1980s, astronomy found itself at a double watershed, if such a thing is possible. On the one hand, advances in technology, inspired by the multiple mirror telescope, had enabled astronomers to contemplate an increase in observing sensitivity that was an order of magnitude greater than they had previously experienced. On the other hand, almost all the traditional observatory sites, especially in the southwestern United States, were either severely affected by light pollution or were too low in elevation to be suitable for infrared and submillimeter observations — or both. Sites falling into the former category include Mt. Palomar, Mt. Hamilton, Mt. Wilson, Mt. Lemmon, and increasingly Mt. Hopkins and Kitt Peak; sites in the latter category include Mt. Palomar, Mt. Hamilton, Mt. Wilson, Mt. Hopkins, Kitt Peak, Mt. Locke, and the existing Chilean sites. Only Mauna Kea, Hawaii, fulfills the necessary atmospheric conditions, but it does so at a substantial penalty in secondary site conditions. It is also rapidly filling up. Clearly, world astronomy needed to establish at least one new site for astronomical observation, one with performance criteria even more demanding than for any previous selection. The problem was especially acute for the United States, which had for almost a century led the field using facilities located in the Southwest at sites that are simply not viable for the next generation of telescopes. However, the continental United States still meets the atmospheric, topographic, and logistical requirements for a good site. Accordingly, astronomers of the Smithsonian Institution and the University of Arizona initiated a search for a new site in 1980.

Selection of Mt. Graham

Clear skies and low water vapor are the key basic criteria in selecting a site. Radio-sonde data (sounding with radio waves to measure atmospheric clarity) provide a clear picture of the global distribution of atmospheric water vapor content, and annual sunshine-hour maps provide a good indication of clear sky conditions. Figure 5.2 shows all mountains above 9,000 feet (2,740 m) in the continental United States superimposed on a map showing annual sunshine-hour contours. This figure suggests that the clearest skies are over Yuma, Arizona, with good conditions extending over much of southern California and southern and western Arizona. When peak elevation is plotted against sunshine-hours (figure 5.3), the best sites within any specified altitude range lie along the upper boundary of this plot.

This criterion led to the selection of the top 24 sites for further study, along with 3 sites that already have major observatories: Kitt Peak, Mt. Hopkins, and Sacramento Peak. For each of these sites (table 5.1), we computed quality indices for clear skies (from the sunshine-hour maps), water vapor (from the altitude and latitude using radio-sonde data), sky coverage (from the latitude), sky darkness (from the distance

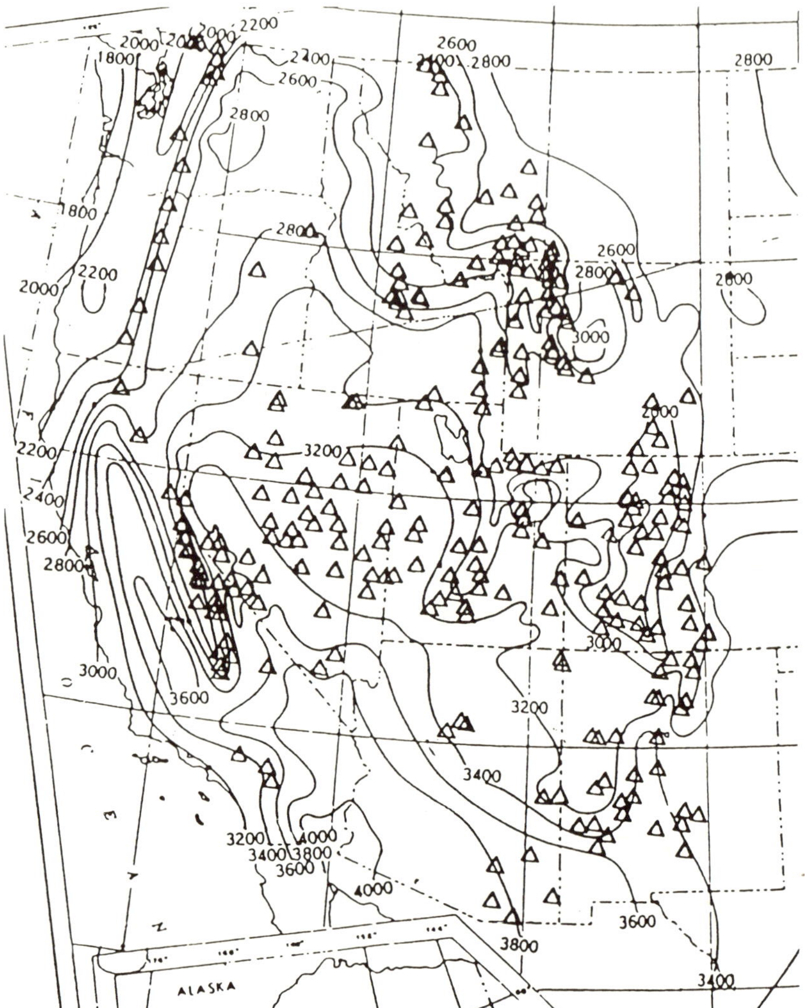

Figure 5.2 Locations of mountains above 9,000 feet (2,750 m) in the western United States. The contours plot sunshine-hours per year, an indication of the typical time free from overcast skies.

to and population of nearby towns), and wind speed (from regional meteorological data). These indices provide a comparative assessment of site quality.

Obviously, the calculated site quality will depend on the weight given to each index, which in turn depends on the scientific program to be undertaken. To illustrate this point we constructed overall indices for three types of astronomy:

1. general optical astronomy (corresponding roughly to the science for which the current generation of 4 m telescopes was built) for which the quality index Q_G is taken as the product of the clear sky, sky coverage, and wind speed indices;
2. infrared/submillimeter astronomy for which the quality index QIR is taken as the product of Q_G and the water vapor index;
3. faint-object astronomy for which the quality index Q_D is taken as the product of Q_G and the sky darkness index.

The results are shown in table 5.2, which lists the sites in order of merit and the corresponding percentage decrement from the best site in each category.

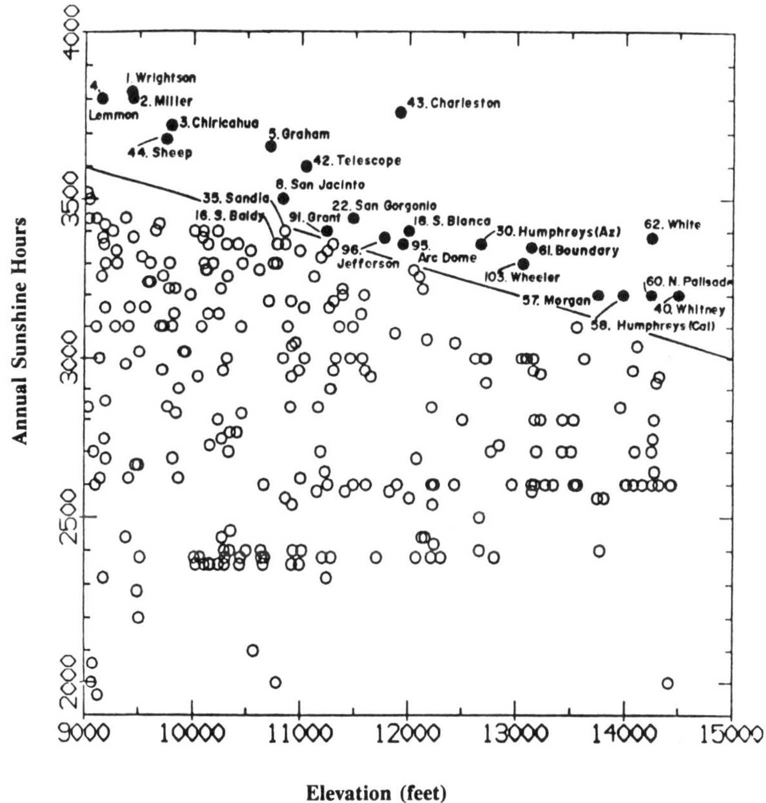

Figure 5.3 Sunshine-hours plotted against altitude for mountains in the continental United States above 9,000 feet (2,750 m). The solid line represents the cutoff used in selecting the basic sample of sites listed in all the tables for this chapter. A few sites below this cutoff were included in the final sample because of special characteristics.

Table 5.1 Analysis of Physical Measurements for 27 Prospective Sites for Development of Astronomical Facilities

Sites	Elevation (feet)	Sunshine-Hours	Latitude	Light Pollution[1]	Clear Sky Index[2]	Sky Darkness Index[3]	Water Vapor Index[4]	Wind Speed Index[5]	Latitude Index[6]
Arc Dome	11,775	3,380	38.8	.00	.738	1.00	.913	.84	.81
Boundary	13,140	3,350	37.8	.02	.724	.94	1.030	.82	.81
Charleston	11,918	3,760	36.3	.22	.928	.60	.878	.88	.83
Chiricahua	9,795	3,720	31.8	.02	.908	.94	.576	1.00	.86
Graham	10,713	3,660	32.7	.04	.877	.89	.685	1.00	.85
Grant	11,239	3,400	38.6	.00	.748	1.00	.856	.82	.81
(Hopkins)	8,550	3,820	31.7	.15	.959	.69	.449	1.00	.86
Humphreys, Az.	12,670	3,360	35.3	.26	.728	.56	.933	.98	.84
Humphreys, Cal.	13,978	3,200	37.3	.02	.651	.94	1.104	.82	.81
Jefferson	11,949	3,360	38.8	.00	.728	1.00	.931	.86	.81
(Kitt)	6,900	3,900	31.9	.12	1.000	.74	.288	.98	.86
Lemmon	9,157	3,800	32.4	1.11	.948	.23	.524	1.00	.86
Miller	9,445	3,800	31.4	.23	.948	.59	.532	1.00	.87
Morgan	13,748	3,200	37.4	.02	.651	.94	1.083	.82	.81

N. Palisade	14,242	3,200	37.1	.02	.651	.94	1.126	.82	.81
Sacramento	9,250	3,420	32.8	.12	.758	.74	.541	1.00	.86
Sandia	10,852	3,400	35.2	.87	.748	.28	.749	1.00	.84
San Gorgonio	11,485	3,440	34.1	.30	.768	.53	.790	.86	.84
San Jacinto	10,831	3,500	33.8	.30	.797	.53	.719	.86	.84
Sheep	9,750	3,680	36.6	.18	.887	.65	.667	.88	.83
Sierra Blanca	12,003	3,400	33.4	.02	.748	.94	.828	1.00	.85
South Baldy	10,787	3,380	34.0	.04	.738	.89	.719	1.00	.84
Telescope	11,047	3,600	36.2	.06	.847	.85	.789	.86	.83
Wheeler	13,063	3,300	39.0	.00	.699	1.00	1.046	.90	.81
White	14,246	3,380	37.6	.02	.738	.94	1.137	.82	.81
Whitney	14,495	3,200	36.6	.05	.651	.87	1.141	.82	.83
Wrightson	9,432	3,820	31.7	.15	.959	.69	.537	1.00	.86

[1]City light contribution at 45° elevation as a percentage of natural sky brightness at visible wavelengths.
[2]Normalized to unity at Kitt Peak.
[3]Computed from formula $d = 1/(1 + 3p)$ where p is the light pollution percentage given in column four; d is approximately the reduction in effective aperture, as described in Burstein et al. 1984.
[4]The water vapor index is essentially the same as that used in Lynds and Goad 1984.
[5]Index computed from maximum recorded wind speed data.
[6]Fraction of sky observable above 20° elevation normalized to value for an equatorial site.

Table 5.2 Ranking of the 27 Sites for General Optical Astronomy, Infrared and Submillimeter Astronomy, and Faint Object Astronomy

Rank	General Optical	ΔQ_G	Infrared/Submillimeter	ΔQ_{IR}	Faint Object	ΔQ_D
	Site	%	Site	%	Site	%
1	Kitt	0.0	Charleston	0.0	Chiricahua	0.0
2	Miller	2.1	Humphreys, Az.	6.0	Graham	9.6
3	Wrightson	2.2	White	6.3	Kitt	15.8
4	Hopkins	2.2	Wheeler	10.3	Sierra Blanca	18.5
5	Lemmon	3.2	Sierra Blanca	11.5	Wrightson	22.8
6	Chiricahua	7.4	Graham	14.1	Hopkins	22.8
7	Graham	11.5	Whitney	14.9	South Baldy	24.8
8	Charleston	19.6	Boundary	16.8	Telescope	30.4
9	Sacramento	22.7	N. Palisade	18.1	Wheeler	30.8
10	Sheep	23.1	Humphreys, Cal.	19.6	Jefferson	31.1
11	Sierra Blanca	24.6	Telescope	19.9	Arc Dome	31.8
12	Sandia	25.5	Jefferson	20.6	Grant	32.5

13	South Baldy	26.4	Sandia	20.9	Miller	33.7
14	Telescope	28.3	Morgan	21.3	Sacramento	34.9
15	Humphreys, Az.	28.8	Arc Dome	22.9	White	37.2
16	San Jacinto	31.7	Chiricahua	24.5	Boundary	38.4
17	San Gorgonio	34.2	South Baldy	25.1	Sheep	42.8
18	Wheeler	39.5	Wrightson	25.6	Humphreys, Cal.	44.5
19	Jefferson	39.8	Miller	26.2	Charleston	44.6
20	Arc Dome	40.4	San Gorgonio	26.3	N. Palisade	44.6
21	Grant	41.1	Sheep	27.3	Morgan	44.6
22	White	41.8	Lemmon	28.2	Whitney	47.6
23	Boundary	43.0	Grant	28.5	Humphreys, Az.	54.2
24	Whitney	47.4	San Jacinto	30.4	San Jacinto	58.8
25	Humphreys, Cal.	48.6	Hopkins	37.8	San Gorgonio	60.4
26	N. Palisade	48.7	Sacramento	40.7	Lemmon	74.4
27	Morgan	48.7	Kitt	59.2	Sandia	76.4

Note: The ΔQs denote the percentage of reduction in quality from the best site in each category.

Although this procedure may not be totally accurate, it nonetheless appears to be objective and to show no obvious inconsistencies. Thus, in the category of general optical astronomy, the existing observatory sites in the desert Southwest—Kitt Peak, Mt. Hopkins, Mt. Lemmon, and Sacramento Peak—all rank in the top ten, along with immediately neighboring sites such as Mt. Wrightson, Miller Peak, Chiricahua Peak, and Mt. Graham. However, only two of the top ten in this category—Charleston Peak and Mt. Graham—rank in the top ten infrared sites. The existing observatory sites all rank very low in this category, as noted above. Finally, for faint-object astronomy, Chiricahua Peak and Mt. Graham place first and second, while Charleston Peak, which is near Las Vegas, appears much lower because of light pollution.

Thus, selecting sites for different types of astronomical research is difficult at best (table 5.2). Since astronomers must accommodate all three disciplines, the question then arises whether to establish three (or at least two) different sites or to identify one site that is adequate for all purposes. The former is the astronomically favored solution but involves much greater cost both in financial and in environmental terms; the latter requires a certain amount of compromise in regard to the astronomical conditions but is clearly more cost effective.

We therefore attempted to assess site quality for the various combinations of astronomical research, giving equal weight to each included discipline (table 5.3). The combined quality index represents some measure of the quality decrement for a combined observatory site compared to the best site for each individual discipline. For the "new astronomy" in which researchers emphasize the study of very faint objects and making infrared and submillimeter observations, Mt. Graham and Chiricahua Peak top the list. When all three disciplines are given equal weight, these two sites reverse order but still offer the best conditions.

Table 5.4 lists, in order, the top 12 sites for overall astronomy (the combination of all three disciplines) along with their ranking for faint-object astronomy and infrared and submillimeter observations, a description of their current use, their availability, and, where relevant, access and cost factors. A number of otherwise excellent sites are excluded because they are in wilderness areas or are otherwise designated as reserved areas. The availability of some other sites is doubtful, and, in any case, these sites have poor access—implying the need to build costly roads and to incur the corresponding environmental damage. (Existing sites, such as Kitt Peak, Mt. Hopkins, etc., are poor candidates for the "new astronomy" because of their poor infrared rating.) The clear exception is Mt. Graham, which ranks among the top two sites.

Mt. Graham Research Site

Mt. Graham is located in the Coronado National Forest and is accessible to 2,700 m in elevation by a paved two-lane highway (Swift Trail; AZ-366). Several peaks above 3,000 m are distributed along a ridge extending from Webb Peak in the northwest to Heliograph Peak in the south (figure 5.4a). All, except Hawk Peak, are accessible by existing dirt roads. We excluded the otherwise very promising Hawk Peak site from further consideration to avoid disturbing this relatively untouched area. Instead we focused on High Peak and Emerald Peak, both of which had already been heavily affected by human activity (see figure 5.4b). Mt. Graham proper (High Peak), which is a relatively small area at the highest elevations, was ideally suited for submillimeter work. The somewhat lower-elevation, more extensive Emerald Peak site promised better image sharpness and was, therefore, more suitable for optical/infrared telescopes and for interferometry.

We felt then, and still do today, that Mt. Graham is not only the best available astronomical site in the continental United States (Steward Observatory 1986), but is also an environmentally sound choice: any impact an observatory may have is small compared to those that have already been incurred on Mt. Graham. Thus, even the most extensive observatory ever proposed for Mt. Graham would have required less than 24 ha—compared to the 4,500 ha that have previously been logged. Also, the total increased use of the mountain by astronomers would be significantly less than the average annual increase in visitor traffic over the last several years. (The Forest Service recorded approximately 250,000 visitor recreation days in 1988 and projected 500,000 visitor recreation days by the second decade of the next century [USFS 1988].) Unless the Swift Trail is closed, this public use will continue to grow.

Subsequent Events

After the original selection of Mt. Graham, we proceeded to assess the one important criterion that we had not yet measured: image sharpness. Such measurements can be made only in situ and are rather time consuming. The early results for Mt. Graham (which have since been

Table 5.3 Rankings of the 27 Sites for Combinations of Astronomical Research

Rank	Faint Object/Infrared Δ		Faint Object/General Optical Δ		Infrared/General Optical Δ		All Three Combined Δ	
	Site	%	Site	%	Site	%	Site	%
1	Graham	11.8	Chiricahua	3.6	Charleston	9.8	Chiricahua	10.6
2	Chiricahua	12.2	Kitt	7.9	Graham	12.8	Graham	11.7
3	Sierra Blanca	15.0	Graham	10.5	Wrightson	13.8	Wrightson	16.8
4	Wheeler	20.5	Wrightson	12.4	Miller	14.1	Sierra Blanca	18.2
5	White	21.7	Hopkins	12.4	Lemmon	15.7	Miller	20.6
6	Charleston	22.3	Miller	17.8	Chiricahua	15.9	Hopkins	20.9
7	Wrightson	24.2	Sierra Blanca	21.5	Humphreys, Az.	17.4	Charleston	21.3
8	South Baldy	24.9	South Baldy	25.6	Sierra Blanca	18.0	Kitt	25.0
9	Telescope	25.1	Sacramento	28.7	Hopkins	19.9	South Baldy	25.4
10	Jefferson	25.8	Telescope	29.3	Sandia	23.1	Telescope	26.1
11	Arc Dome	27.3	Charleston	32.0	Telescope	24.0	Wheeler	26.8
12	Boundary	27.6	Sheep	32.9	White	24.1	White	28.4

13	Miller	29.9	Wheeler	35.1	Wheeler	24.9	Humphreys, Az.	29.7
14	Humphreys, Az.	30.1	Jefferson	35.4	Sheep	25.2	Jefferson	30.5
15	Hopkins	30.3	Arc Dome	36.1	South Baldy	25.7	Sheep	31.1
16	Grant	30.5	Grant	36.7	Kitt	29.6	Arc Dome	31.7
17	Whitney	31.3	Lemmon	38.8	Boundary	29.9	Boundary	32.7
18	N. Palisade	31.3	White	39.5	Jefferson	30.2	Sacramento	32.7
19	Humphreys, Cal.	32.0	Boundary	40.7	San Gorgonio	30.2	Grant	34.0
20	Morgan	32.9	Humphreys, Az.	41.5	San Jacinto	31.0	Lemmon	35.2
21	Sheep	35.1	San Jacinto	45.2	Whitney	31.1	Whitney	36.6
22	Kitt	37.5	Humphreys, Cal.	46.5	Arc Dome	31.6	N. Palisade	37.1
23	Sacramento	37.8	N. Palisade	46.6	Sacramento	31.7	Humphreys, Cal.	37.5
24	San Gorgonio	43.3	Morgan	46.6	N. Palisade	33.4	Morgan	38.1
25	San Jacinto	44.6	Whitney	47.2	Humphreys, Cal.	34.1	San Gorgonio	40.3
26	Sandia	51.3	San Gorgonio	47.5	Grant	34.8	San Jacinto	40.3
27	Lemmon	51.3	Sandia	50.9	Morgan	34.9	Sandia	40.9

Note: The Δs represent the normalized sum of relevent individual ΔQs listed in table 5.2.

Table 5.4 Leading Prospective Sites for Development of Astronomical Facilities

Combined Ranking	Mountain	Faint Object/ Infrared Ranking	Present Use	Availability	Access	Cost Factors
1	Chiricahua	2	wilderness area	E	—	—
2	Graham	1	national forest	A	G	G
3	Wrightson	7	wilderness area	E	—	—
4	Sierra Blanca	3	ski resort	D	P	P
5	Miller	13	wilderness area	E	—	—
6	Hopkins	15	observatory	A	F/G	G
7	Charleston	6	wilderness study area	D	P	P
8	Kitt	22	observatory	A	G	VG
9	South Baldy	8	lightning research station	A	F	F
10	Telescope	9	national monument	E	—	—
11	Wheeler	4	national scenic area	E	—	—
12	White	5	former observatory	A	F	P

Note: E = excluded; A = available; D = doubtful; VG = very good; G = good; F = fair; P = poor.

confirmed in great detail) were very encouraging and led, in 1984, to the University of Arizona proposal to the Forest Service to establish an international observatory on the mountain.

By 1988, the Office of Arid Land Studies of the university and the Forest Service had each conducted a biological survey as preparation for the required environmental impact statement. The Mt. Graham red squirrel had been proposed for listing as endangered by the Arizona Game and Fish Department in 1985 and had been so listed by the Fish and Wildlife Service in 1986. On behalf of the Forest Service, biologists prepared two biological assessments for two different observatory proposals (Coronado National Forest 1988), and the Fish and Wildlife Service drafted two corresponding biological opinions, the second being officially issued in July 1988 (Spear 1987, 1988). Forest Service and Fish and Wildlife Service administrators had held four public hearings, but no one has been able to demonstrate quantitatively (then or since) that the observatory project would significantly harm the environment or the squirrel population. The university accepted the provisions of the third "reasonable and prudent alternative" of the final biological opinion, which allowed the construction of an observatory only at Emerald Peak and required, among other conditions, the construction of a new road and the closure of existing ones.

The research area consists of 60 ha, and the area directly affected by the initial telescope sites comprises 3.5 ha including the new road (figure 5.4a). Figure 5.4b is an aerial photograph taken in 1991 at the same scale as figure 5.4a, showing some of the previously logged areas as marked on figure 5.4a.

Astronomers had the new 3 km access road from Swift Trail to the site constructed in a position that would avoid close proximity to any squirrel middens. For most of its length, this plan presented no problem because the area had largely been cleared by previous logging activities. Construction of the first two telescopes and a utility building has been completed; site preparation for the third telescope has begun—but further progress is currently (December 1994) being delayed by litigation.

The Initial Telescopes

The first three telescopes to be placed at Emerald Peak are the Max Planck/Arizona Submillimeter Telescope (SMT), the Vatican Advanced Technology Telescope (VATT), and the Large Binocular Telescope (LBT—formerly known as the Columbus Project Telescope).

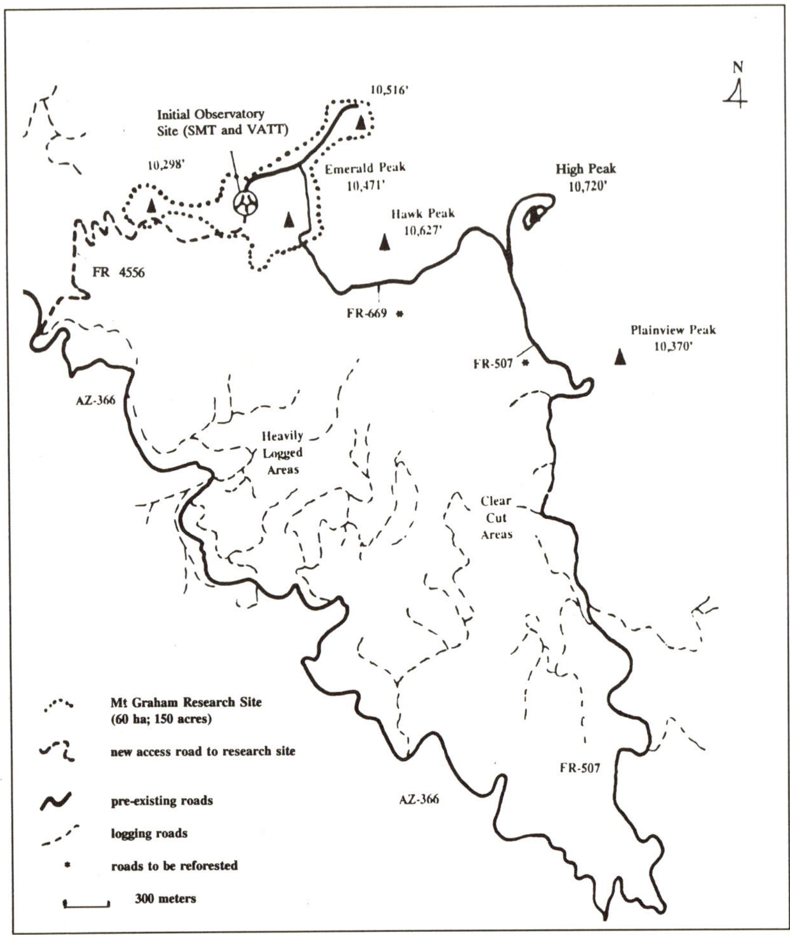

Figure 5.4a Map of the region of the Pinaleño Mountains surrounding the location of the Mt. Graham International Observatory on Emerald Peak.

The SMT is now named after Heinrich Hertz, the German physicist who first demonstrated the electromagnetic nature of light; the enclosure is named after physician Eugene Frazie of Mesa, Arizona. It is a 10 m telescope designed specifically for observations at wavelengths between 0.3 and 1.3 mm. In this region of the spectrum, cool interstellar clouds emit most of their radiation and, hence, most of the information we can ever hope to gather from them. These cool, dusty, molecular clouds are a relatively recent discovery and appear to be sites of current or incipient star formation. If we are ever to understand how

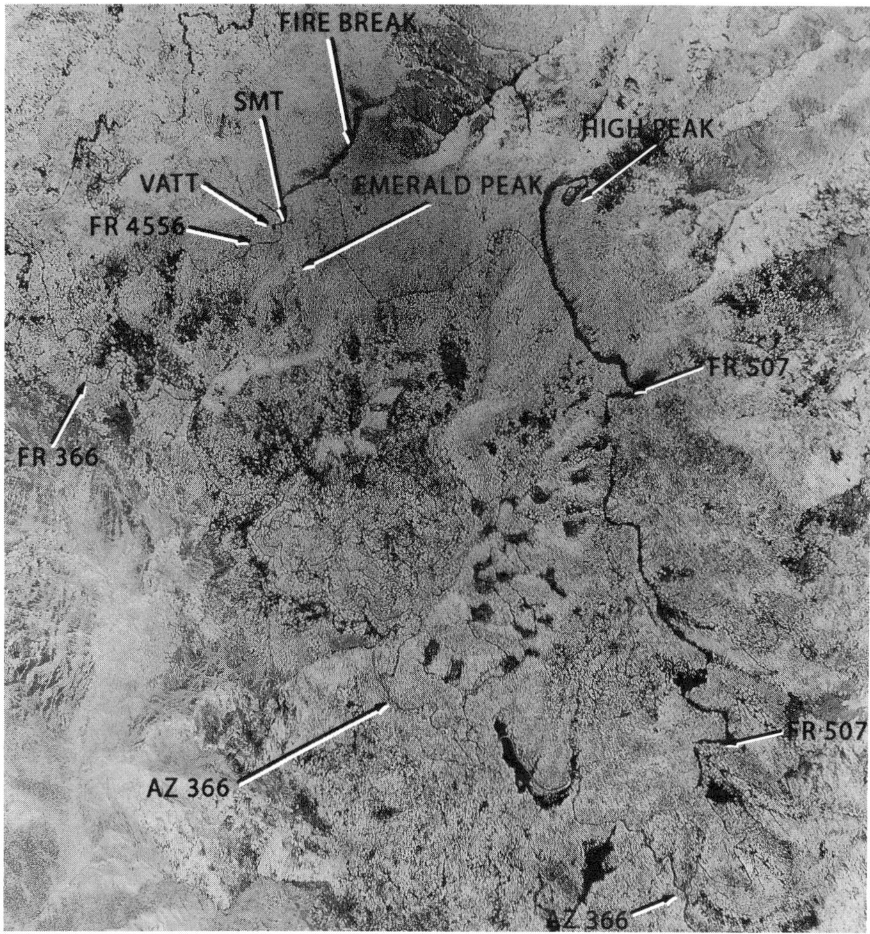

Figure 5.4b Aerial photograph (negative) of the same area mapped in Figure 5.4a. The photograph was taken during the winter of 1991, hence the open areas appear as dark due to snow cover. The area covered by the photograph is approximately 8000 acres surrounding Emerald Peak. Observatory construction includes clearings for the VATT (0.3 acres), the SMT (0.4 acres), and the two-mile observatory access road (FR4556, 1.2 acres). Snow cover (dark areas) also delineates areas of higher elevation and of pre-observatory impacts; noteworthy among these are state highway AZ366, Forest Service road FR507 and the High Peak (Mt. Graham) fire break, the fire break at Emerald Peak, and extensive areas of recent timber harvesting between AZ366, FR507, and FR4556. Lighter areas are those that remain forested. In 1993 a third clearing for the LBT (1.2 acres) was made close to the point labeled as Emerald Peak.

stars begin to form and how pre-planetary disks develop, we must be able to observe the interior structure and dynamics of such clouds. They are, however, completely opaque at visible wavelengths and are observable in the infrared region only after stars have been through much of the formation process. The key to understanding the formation process lies with millimeter and submillimeter observations. The submillimeter region is also critical to understanding the emission mechanisms in objects such as quasars and other very compact radio sources, which are probably associated with the formation of giant black holes in the nuclei of galaxies.

The SMT represents a true breakthrough in the design of radio telescopes with ultra-high surface accuracy, in this case maintaining a surface accuracy of 17 μ over the entire 10 m surface. The SMT achieves this accuracy through the use of composite materials (carbon fiber, reinforced plastic) that have a strength-to-weight ratio 5 times that of steel and a coefficient of thermal expansion the same as low-expansion glass. Use of this material results in a structure that is exceedingly stable, both mechanically and thermally. The SMT, which is a joint project between the Max Planck Institute for Radioastronomy in Bonn, Germany, and the University of Arizona, is depicted in figure 5.5. It was constructed partly in Germany, partly in Tucson.

The Vatican Advanced Technology Telescope (VATT), dedicated to Alice Lennon and housed in the Bannan Astrophysical Facility, is the prototype for the new generation of large optical telescopes in that it has a very "fast" focal ratio (f/1) primary mirror made from a spin-cast borosilicate honeycomb blank. The Steward Observatory Mirror Laboratory, located beneath the University of Arizona's football stadium, cast and polished this 1.8 m mirror. The optics represent a major breakthrough in mirror technology and will permit the construction of more compact (and hence higher-performance and lower-cost) telescopes to be constructed in the future. This technology is also key to the LBT, as well as several other very large instruments. The optical configuration, which optical engineers only a few years ago viewed as technically impossible, also permits smaller secondary mirrors and hence better performance in the optical and thermal infrared regions.

The telescope is a collaboration between the Vatican Observatory in Castelgandolfo, Italy, and the University of Arizona. Astronomers will use it for optical and near-infrared imaging, optical spectroscopy, and optical polarimetry; they may also use VATT for mapping at thermal infrared wavelengths (10 μ), exploiting the excellent conditions at

Figure 5.5 Sketch of the Max Planck/Arizona Submillimeter Telescope (SMT) and its enclosure. This telescope and the Vatican Advanced Technology Telescope (VATT) a short distance away are now in operation.

Emerald Peak. The astronomical program will include studies of the magnetic fields in regions of star formation, the later stages of the star formation process, and the optical and infrared counterparts of radio and X-ray sources.

Construction and operation of the LBT is a collaboration between the Arcetri Observatory of Florence, Italy, the Research Corporation of Tucson, Arizona, and the University of Arizona; astronomers from Ohio State University will also have observing rights at the telescope. The LBT will consist of two f/1.14, 8.4 m mirrors (of spin-cast borosilicate to be made at the Steward Observatory Mirror Laboratory) on a common mount (figure 5.6). The instrument builds substantially on concepts pioneered at the Multiple Mirror Telescope on Mt. Hopkins. The LBT will have the largest collecting area (11.8 m equivalent aperture) of any of the telescopes that are currently planned. It will also

Figure 5.6 Sketch of the Large Binocular Telescope (formerly referred to as the Columbus Project) with its two 8.4 m mirrors. This instrument, now ready for construction, will have a collecting area equivalent to an aperture of 11.8 m.

have by far the longest baseline (22 m) and, hence, the greatest image sharpness capability of any single telescope. It promises to provide a major advance in optical and mechanical performance, as well as in overall performance per unit cost.

Astronomers will use the LBT for a wide variety of research projects, including the study of galaxy formation at epochs around 10–15 billion years ago, determination of the evolution of chemical abundances over this time span, high-resolution imaging and spectroscopy of quasars and other galactic nuclei, analysis of the later stages of star and planet formation, searches for planetary companions to other stars, and imaging of the surfaces of stars.

Conclusions

From the astronomical point of view, an evaluation of the situation on Emerald Peak is as follows. Worldwide, astronomers need new sites for ground-based astronomical research, as most of the existing sites either are degraded by light pollution or are too low in elevation to permit infrared or submillimeter observations. The southwestern United States offers excellent conditions for astronomical research, and Emerald Peak is not only among the best but is also one of the few sites that are even potentially available.

The Mt. Graham International Observatory will encompass some of the world's most advanced telescopes, designed to make major breakthroughs in our understanding of the origin and evolution of the universe over the approximately 15 billion years since its creation in the so-called Big Bang. The observatory is also host to biological research teams studying the red squirrel and other unique ecological features of the site. The Mt. Graham observatory project is thus destined to continue the tradition of close collaboration between astronomers and biologists that was established by Andrew Ellicott Douglass, the founding director of both the Steward Observatory and the Arizona Tree-Ring Laboratory. I hope we can ensure that this tradition continues to flourish in the future.

References

Burstein, D., T. D. Hogan, and M. D. Kroelinge. 1984. *Astronomy and Communities.* Arizona State University, Tempe.

Coronado National Forest. 1988. *Mount Graham Red Squirrel: An Expanded Biological Assessment.* Coronado National Forest, Tucson, Ariz.

Lynds, C. R., and J. W. Goad. 1984. *Publications of the Astronomical Society of the Pacific* 96:583.

Spear, M. J. 1987. *Draft Biological Opinion.* U.S. Department of the Interior, Fish and Wildlife Service, Albuquerque, N.M.

Spear, M. J. 1988. *Biological Opinion.* U.S. Department of the Interior, Fish and Wildlife Service, Albuquerque, N.M.

Steward Observatory. 1986. "The Scientific Justification for the Mt. Graham International Observatory." Submitted to the U.S. Forest Service in March 1986. (Unpublished.)

USFS. 1988. *Final Environmental Impact Statement, Proposed Mt. Graham Astrophysical Area, Pinaleño Mountains, Coronado National Forest.* U.S. Department of Agriculture, Forest Service, Southwestern Region, Tucson, Ariz.

3

The Forests of the Pinaleños

Structure and History

The distribution and successional patterns of the western montane [conifer] species are most obviously coupled to elevational effects such as the substantial decrease in temperature that occurs with increasing elevation, along with numerous other biophysical factors. This biophysical coupling is reflected in the often sharp demarcations between the dominant overstory conifers. A strong influence by the physical environment in combination with the occurrence of fire, results in a complex successional system. In many areas, disturbance appears to be followed by a relatively unchanging physiognomy whereas other areas show a more structured successional sequence of understory-overstory replacements.

W. K. SMITH, in *Physiological Ecology of North American Plant Communities*

The two chapters in Part 3 focus on the high-elevation forests of the Pinaleño Mountains. The questions asked, data presented, conclusions drawn, and discussions identify many of the most critical issues for the future of these forests. The chapter by Juliet Stromberg and Duncan Patten provides a detailed analysis of species composition and stand structure in the spruce-fir forests, as well as their succession, regeneration, and cone production, to lay a sound basis for future plant-ecological studies in the Pinaleños. The chapter by Henri Grissino-Mayer and Harold Fritts employs dendrochronology to develop a refined analysis of past ecological effects in Pinaleño forests and other forests of the region—an essential time sense. They read the fire history of the Pinaleño forests from the tree rings. Extrapolating from their dendrochronological studies, they provide a basis for assessing a variety of environmental effects on trees in the future, including those that may come from air pollution and climatic warming.

The Pinaleño forests are essential for protection of the water resources of the mountain range and beyond, especially the picturesque, high-elevation springs, cienegas, and streams, as well as the beautiful Aravaipa Canyon and stream to the west—and the aquatic life these waters nurture. We must use this wealth of information in planning future protection and use of the entire ecological system of the Pinaleños, as well as for establishing strategies to perpetuate the red squirrel population and the many other plant and animal populations of its ecological community.

CHAPTER SIX

Vegetation Dynamics of the Spruce-Fir Forests of the Pinaleño Mountains

Juliet C. Stromberg and Duncan T. Patten

The Pinaleño Mountains in southeastern Arizona rise over 2,000 m (6,600 feet), from a base elevation of 1,200 m to peaks of 3,270 m. This elevation gradient spans many vegetation zones — desertscrub near the base, juniper-oak woodland and ponderosa pine forest at middle elevations, mixed-conifer forests at upper elevations, and Engelmann spruce (*Picea engelmannii*) and corkbark fir (*Abies lasiocarpa* var. *arizonica,* now *A. bifolia*) forests at the high subalpine zone (> 3,000 m) (Johnson 1988).

Montane spruce-fir forests are widely distributed throughout the Rocky Mountain floristic region, from Canada to Mexico (Peet 1988). However, those on the Pinaleño Mountains are unique for several reasons. First, they exist primarily as almost pristine, old-growth stands, an increasingly rare forest condition in the western United States (Whitney 1987; Habeck 1988; Spies and Franklin 1988; Kaufmann et al. 1992). Second, though limited in area, they represent the only well-developed occurrence of this forest type in southern Arizona. Other mountain "sky islands" in southern Arizona are not as tall and support only small pockets of fir or spruce trees, or have species-depauperate spruce-fir forests in areas of marginal elevation (Whittaker and Niering 1965; Gehlbach 1981). Finally, because the sky-island forests have been isolated from other montane systems in the Southwest for over 11,000 years, many species, possibly including Engelmann spruce, have diverged from races on nearby mountains (USFS 1988). Corkbark fir also may have diverged from races in the central area of its distribution to the north, the variety in the Pinaleños being in the Madrean, or southernmost, of the Rocky Mountain floristic regions.

A mosaic of patch types occurs within the spruce-fir forest matrix on the Pinaleño Mountains. The landscape is analogous to Swiss cheese, wherein the old-growth forests are punctuated by dry meadows and early successional forests recovering from fire or windthrow and by rare spring-fed marshes (cienegas) at the headwaters of mountain streams. Below about 3,000 m, Engelmann spruce and corkbark fir give way to a mixed-conifer forest of Douglas-fir (*Pseudotsuga menziesii* var. *glauca*), Southwestern white pine (*Pinus strobiformis*), and white fir (*Abies concolor*). Stringers of spruce-fir forest extend downward into this zone along mountain streams, favored by the cool, wet microclimate of the riparian zone. Human activities (e.g., logging) have disturbed much of the mixed-conifer forest and the transitional forest between the spruce-fir and mixed-conifer zones (USFS 1988). The spruce-fir forest has had less human impact, except for recent disturbances (e.g., road construction, tree removal) associated with fire breaks, roads, and observatory construction.

In our study, conducted in 1988, 1990, and 1991, we examined spatial variation in composition, structure, and reproductive output of the spruce-fir forests on the Pinaleño Mountains. Because of the importance of determining the effect of the observatory construction on red squirrel habitat, we were particularly interested in factors associated with temporal change, or postdisturbance redevelopment, in forest composition and structure. Studies in other regions have revealed general trends of spruce-fir vegetation dynamics but also have emphasized the importance of site-specific studies (Oosting and Reed 1952; Merkle 1954; Day 1972; Dye and Moir 1977; Whipple and Dix 1979; Aplet et al. 1988). Studies of this type are particularly important for forests such as those on the Pinaleño Mountains that are disjunct from the main range of the spruce-fir forests. The need to understand the dynamics of forest recovery in response to natural and anthropogenic disturbance is acute, because the old-growth forests of the Pinaleño Mountains are increasingly threatened by human activities.

Methodology

We collected data from study sites within the subalpine zone of the Pinaleño Mountains, primarily on High Peak (Mt. Graham), Hawk Peak, Emerald Peak, and Plain View Peak (Stromberg and Patten 1991 and 1993). Our study sites encompassed forest stands recently disturbed by fire, windthrow, and roadside clearing; stands midway through the recovery process; and old-growth stands on sites varying in slope, rock cover, and elevation. We sampled vegetation plots within the stands for woody plant density and basal area, by species, and for canopy cover. We also collected increment cores from representative trees of varying age within each stand and processed these to determine tree age. In addition, we quantified the reproductive output of Engelmann spruce during a year (1990) of large seed crop (mast year) for the species using direct counts of cones with binoculars and counts of seeds within seed-traps placed on the forest floor. We then compared age structure, stand density, basal area, and reproductive output among stands in varying stages of postdisturbance recovery and among various old-growth forest stands, including those with and without red squirrel (*Tamiasciurus hudsonicus grahamensis*) middens.

Forest Composition and Structure

Spatial Variation

Most of the forest stands in the spruce-fir zone of the Pinaleño Mountains study area qualify for old-growth status based on attributes including a high density of old trees (e.g., more than 20 trees/ha older than 200 years), an abundance of fallen logs, multiple-age cohorts, and high, but patchy, canopy cover (Morrison 1988; Kaufmann et al. 1992). The age of Engelmann spruce ranged among stands from 300 to 380 years, somewhat younger than its maximum age of 500 to 600 years (Moir 1992). Engelmann spruce dominated the large-tree class (those with trunk diameter > 25 cm), with an average of about three large spruce trees for every large corkbark fir tree, whereas most of the smaller trees and saplings were corkbark fir. Engelmann spruce and corkbark fir shared dominance with Douglas-fir in the old-growth forests within the spruce-fir–mixed-conifer transition zone. Shrubs generally covered less than 10% of the forest floor, with blueberry (*Vaccinium oreophilum*) being the dominant shrub. Herbaceous plants, including side-bells pyrola (*Pyrola secunda*), western sedge (*Carex occidentalis*), and strawberry (*Fragaria ovalis*), covered less than 5% of the forest floor.

Stand density, basal area, and abundance of spruce and fir trees varied between sites. Tree density was lower on sloping sites, which occurred mainly at lower elevations in the study area. Forests in the transition zone had an average slope of 35° ± 7°, compared to 15° ± 7° for the spruce-fir zone, and had about half as many trees per unit area (120 vs. 200 per 0.1 ha). In the spruce-fir zone, stand basal area ranged among old-growth sites from about 70 to 90 m^2/ha, and spruce basal area ranged from about 30 to 80 m^2/ha. Spruce basal area was greatest along the edges of wet meadows (cienegas), at sites with low rock cover (which ranged among sites from 0% to 26% cover), and at sites with level terrain. Sites with a high basal area of spruce also tended to be sites with red squirrel middens. For example, spruce basal area averaged 71 ± 20 m^2/ha in forest sites with middens as opposed to 37 ± 15 m^2/ha for stands of similar age without middens.

Temporal Variation

Two general pathways of postdisturbance forest recovery to old-growth status were apparent in the spruce-fir forest zone (figure 6.1). The first type occurred in response to total clearing of the forest, such as after fire. This pathway was characterized by the initial colonization of Engelmann spruce in open meadow sites. Some corkbark fir were present, but spruce saplings outnumbered fir saplings in the young stands by about five to one. The "pioneer" spruce trees had very high growth rates in the open sunny conditions and attained sizes associated with old-growth conditions (i.e., trunk diameters of 50 cm) in about 100 years. However, a much longer time was required for the forest stand to attain other attributes associated with old-growth status (figure 6.2a). About 80 to 150 years after the initial "wave" of spruce colonization, a second wave of spruce establishment occurred, resulting in a bimodal age structure for spruce; the same situation has also occurred in spruce-fir forests in other regions (Rebertus et al. 1992). This second episode of spruce establishment was followed soon after by large-scale establishment of corkbark fir, facilitated by the spruce overstory. The spruce overstory altered the forest microclimate by moderating temperature extremes and decreasing solar irradiation and also modified the forest floor by depositing a thick litter layer that reduced competitive herbaceous cover and increased soil moisture. As the forest matured over time, conditions continued to favor fir. After about 300 years, the forests had about three fir saplings for every spruce sapling. Whereas attributes of fir seeds and seedlings allowed them to establish in shaded sites with deep soil litter, small openings produced by localized tree-fall may have facilitated the continued establishment of spruce in the old-growth forest (Patten 1963; Knapp and Smith 1982; Shea 1985; Veblen 1986). Growth rates of the understory trees slowed as between-tree competition increased and light availability decreased. For example,

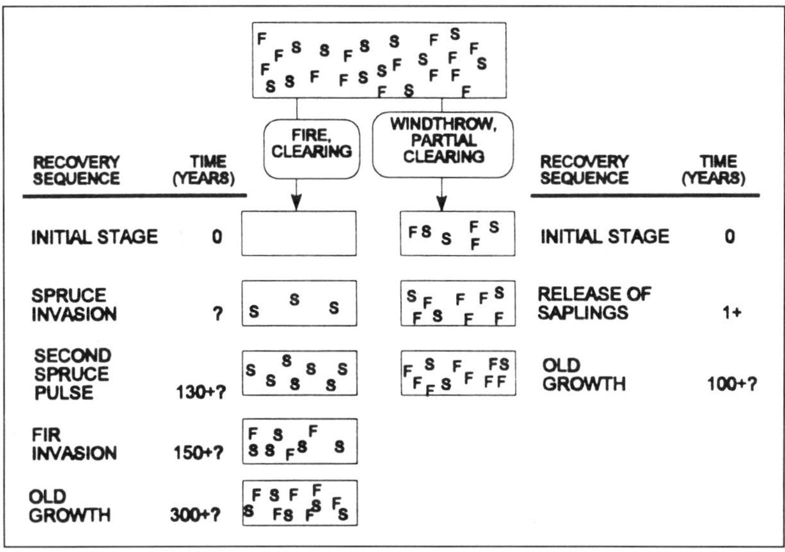

Figure 6.1 Diagrammatic pathway of Engelmann spruce–corkbark fir forest recovery after two types of disturbance. "S" denotes spruce trees, and "F" denotes fir trees.

200 to 250 years were required for understory trees to attain trunk diameters of 20 cm. The initial spruce pioneers died after about 300 to 400 years, remaining as downed or standing dead "snag trees" identifiable as open-grown trees by large branches that extended down the trunk nearly to the forest floor.

Another pattern of postdisturbance recovery occurred after partial stand clearing, such as windthrow. In contrast to the spruce-dominated pathway that occurred after complete stand clearing, this pathway was characterized by corkbark fir establishment. After windthrow, stands in the early stages of recovery were dominated by saplings and small trees that had been released from suppression by removal of the overstory trees. Because corkbark fir dominated the understory of most old-growth forests, it became the dominant tree along windthrows and forest edges.

Forest recovery after stand clearing in the transition zone differed from that in the spruce-fir zone (figure 6.2b). Engelmann spruce and Douglas-fir both functioned as pioneering species in the transition zone, and stand development was slower. Douglas-fir was the initial pioneer in some stands, followed 100 to 300 years later by the estab-

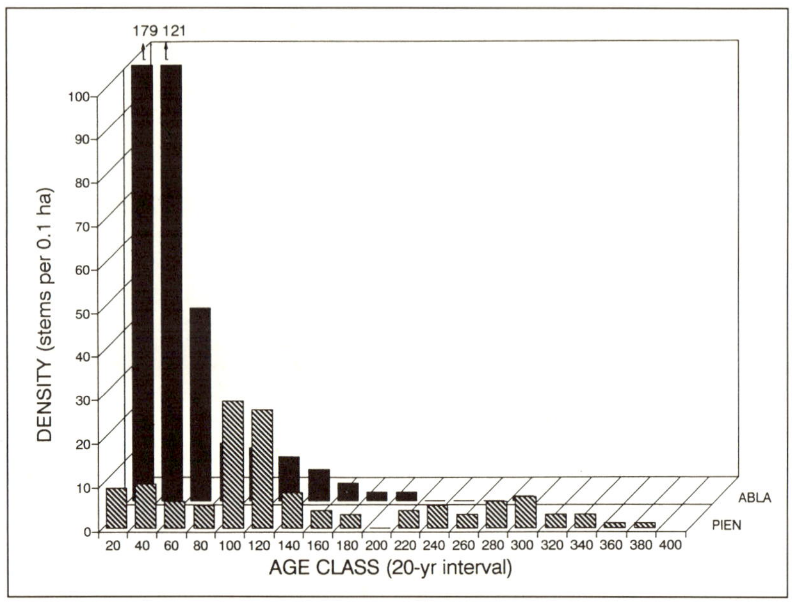

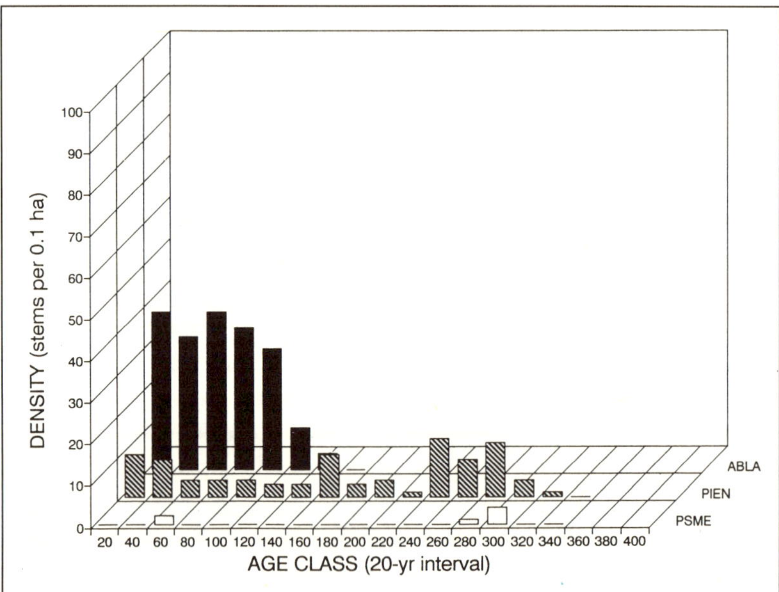

Figure 6.2 Density distribution by age class for *Abies lasiocarpa* (ABLA), *Picea engelmannii* (PIEN), and *Pseudotsuga menziesii* (PSME) at old-growth stands in the spruce-fir zone (figure 6.2a) and spruce-fir mixed-conifer transition zone (figure 6.2b) of the Pinaleño Mountains.

lishment of Engelmann spruce. In other stands, Engelmann spruce and Douglas-fir were co-pioneers. In all cases, corkbark fir did not establish until 150 to 250 years after initial colonization of the pioneer trees, perhaps because amelioration of the stand microclimate required more time in this ecologically "harsher" zone. Douglas-fir, in contrast to Engelmann spruce and corkbark fir, did not continue to establish in the forest understory. Old-growth forests (260 to 385 years) in this zone were dominated by Engelmann spruce and corkbark fir, with Douglas-fir represented by several large senescent or dead trees.

Reproductive Output of Engelmann Spruce

Temporal Variation

Engelmann spruce and corkbark fir are both species that produce large, or mast, seed crops at characteristic intervals of about three to five years. This temporal fluctuation in the size of seed crops significantly influences the annual population size of animal species such as the red squirrel. After producing small seed crops in the prior three years, Engelmann spruce in the Pinaleño Mountains produced a large seed crop in 1990 but a small seed crop again in 1991. Seed production in 1990 in the old-growth forest stands averaged about 25 million seeds/ha and was in the highest (> 1,235,000 seeds/ha) of six seed production classes described in Alexander and Shepperd 1984. Values for cone production per tree were high as well (> 2,000 cones for the largest and oldest trees). Seedfall data represent values harvested as of September 1991 and may underestimate total seed production values because some of the 1990 cones remained unopened on trees even in late 1991.

Seed and cone production by Engelmann spruce during the mast year increased linearly with stand age, in other words, with successional stage of the forest. Cone production per tree was higher in the open, early successional forest, but total cone and seed production by the stand was several times higher in the denser, old-growth forests. For example, forests less than 50 years old produced about 7 million seeds/ha compared to about 21 million seeds/ha for forests older than 230 years.

Spatial Variation

Seed production ranged among old-growth spruce-fir sites from 12 to 49 million seeds/ha. The greatest seed production occurred at sites with low rock cover on level terrain (and presumably with deep soils), which

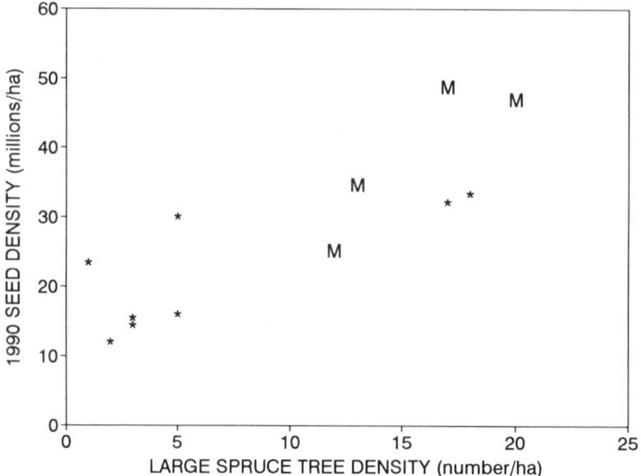

Figure 6.3 Number of *Picea engelmannii* seeds on the forest floor in relation to density of large (> 50 cm trunk diameter) *Picea engelmannii* trees at the site in 1990. "M" denotes forests with red squirrel middens.

supported a high basal area of spruce and exhibited a high density of large (> 50 cm diameter) spruce trees. Large variation in spruce seed and cone production between sites, and the correlation of reproductive output with spruce basal area and density of large trees, also has been observed in the central Rocky Mountains (Roe 1967; Alexander et al. 1986). However, high reproductive output at certain forest sites was not due simply to the greater abundance of seed-producing trees, but also to greater per-tree productivity. For example, sites with red squirrel middens had nearly triple the spruce seedfall of nonmidden forests, an average of 36 million seeds/ha as opposed to 14 million seeds/ha (figure 6.3). This difference was a result, in part, of greater spruce basal area (71 m²/ha vs. 37 m²/ha), but also of higher per-tree cone production: spruce trees at midden-forests produced about 820 cones per tree as opposed to 400 per tree at sites without middens (values are for trees with trunks 50 cm in diameter).

Implications for Forest Management

The mountaintops of the Pinaleños support a diverse mixture of seral forest stages (or forests in different stages of development after disturbance) and of old-growth forest stands with subtle differences in structure, composition, and productivity. Some stands were in the initial stages of recovery from natural disturbance and might attain old-growth status—with its attendant presence of multiple age classes, high canopy cover with localized light gaps, and an abundance of old trees and downed logs—in another 300 to 400 years (Kaufmann et al. 1992). In the interim period, the early successional forests provide important habitat for plant and animal species associated with open meadow or savanna-type conditions (Johnson 1988). Among the old-growth forest stands, those that were near wet meadows or on level terrain with little rock cover had greater productivity of vegetative traits (e.g., stand basal area and tree density) and reproductive traits (e.g., cone and seed production), as has been reported in other studies (Fralish 1994). Thus, within the old-growth forest matrix were "islands" of productivity and seral development. Different areas within the old-growth forest had different "value" as habitat for animal species, including the Mt. Graham red squirrel. This diversity of stand types suggests that management should be guided by a thorough understanding of the temporal dynamics of the forest ecosystem and of the distribution and causes of patch diversity.

References

Alexander, R. R. 1986. Engelmann spruce seed production and dispersal, and seedling establishment in the central Rocky Mountains. General Technical Report RM-134, pp. 1–9. U.S. Department of Agriculture, Forest Service, Ft. Collins, Colo.

Alexander, R. R., and W. D. Shepperd. 1984. Silvical characteristics of Engelmann spruce. General Technical Report RM-114, pp. 1–38. U.S. Department of Agriculture, Forest Service, Ft. Collins, Colo.

Aplet, G. H., R. D. Laven, and F. W. Smith. 1988. Patterns of community dynamics in Colorado Engelmann spruce–subalpine fir forests. *Ecology* 69:312–19.

Day, R. J. 1972. Stand structure, succession, and use of southern Alberta's Rocky Mountain forest. *Ecology* 53:472–78.

Dye, A. J., and W. H. Moir. 1977. Spruce-fir forest at its southern distribution in the Rocky Mountains, New Mexico. *American Midland Naturalist* 97: 133–46.

Fralish, J. S. 1994. The effect of site environment on forest productivity in the Illinois Shawnee Hills. *Ecological Applications* 4:134–43.

Gehlbach, F. R. 1981. *Mountain Islands and Desert Seas: A Natural History of the U.S.-Mexican Borderlands.* Texas A & M University Press, College Station.

Habeck, J. R. 1988. Old-growth forests in the northern Rocky Mountains. *Natural Areas Journal* 8:202–11.

Johnson, W. T. 1988. Flora of the Pinaleño Mountains, Graham County, Arizona. *Desert Plants* 8:147–62, 175–91.

Kaufmann, M. R., W. H. Moir, and W. W. Covington. 1992. Old-growth forests: What do we know about their ecology and management in the Southwest and Rocky Mountain regions? General Technical Report RM-213, pp. 1–11. U.S. Department of Agriculture, Forest Service, Ft. Collins, Colo.

Knapp, A. K., and W. K. Smith. 1982. Factors influencing under story seedling establishment of Engelmann spruce (*Picea engelmannii*) and subalpine fir (*Abies lasiocarpa*) in southeast Wyoming. *Canadian Journal of Botany* 60: 2753–61.

Merkle, J. 1954. An analysis of the spruce-fir community on the Kaibab Plateau, Arizona. *Ecology* 35:316–22.

Moir, W. H. 1992. Ecological concepts in old-growth forest definition. General Technical Report RM-213, pp. 18–23. U.S. Department of Agriculture, Forest Service, Ft. Collins, Colo.

Morrison, P. H. 1988. *Old-Growth in the Pacific Northwest: A Status Report.* The Wilderness Society, Washington, D.C.

Oosting, H. J., and J. F. Reed. 1952. Virgin spruce-fir forest of the Medicine Bow Mountains, Wyoming. *Ecological Monographs* 22:69–91.

Patten, D. T. 1963. Light and temperature influence on Engelmann spruce seed germination and subalpine forest advance. *Ecology* 44:817–18.

Peet, R. K. 1988. Forests of the Rocky Mountains. In *North American Terrestrial Vegetation,* ed. M. G. Barbour and W. D. Billings, pp. 63–101. Cambridge University Press, New York.

Rebertus, A. J., T. T. Veblen, L. M. Roovers, and J. N. Mast. 1992. Structure and dynamics of old-growth Engelmann spruce–subalpine fir in Colorado. General Technical Report RM-213, pp. 139–53. U.S. Department of Agriculture, Forest Service, Ft. Collins, Colo.

Roe, A. L. 1967. Seed dispersal in a bumper spruce seed year. General Technical Report INT-39, pp. 1–10. U.S. Department of Agriculture, Forest Service, Ogden, Utah.

Shea, K. L. 1985. Demographic aspects of coexistence in Engelmann spruce and subalpine fir. *American Journal of Botany* 72:1823–33.

Spies, T. A., and J. F. Franklin. 1988. Old growth and forest dynamics in the Douglas-fir region of western Oregon and Washington. *Natural Areas Journal* 8:190–201.

Stromberg, J. C., and D. T. Patten. 1991. Dynamics of the spruce-fir forests on the Pinaleño Mountains, Graham County, Arizona. *Southwestern Naturalist* 36:37–48.

Stromberg, J. C., and D. T. Patten. 1993. Seed and cone production by Engelmann spruce in the Pinaleño Mountains, Arizona. *Journal of the Arizona-Nevada Academy of Science* 27:79–88.

U.S. Forest Service. 1988. *Final Environmental Impact Statement, Proposed Mt. Graham Astrophysical Area, Pinaleño Mountains, Coronado National Forest.* EIS #03-05-88-1. U.S. Department of Agriculture, Forest Service, Tucson, Ariz.

Veblen, T. T. 1986. Tree falls and the coexistence of conifers in subalpine forests of the central Rockies. *Ecology* 67:644–49.

Whipple, S. A., and R. L. Dix. 1979. Age structure and successional dynamics of a Colorado subalpine forest. *American Midland Naturalist* 101:142–58.

Whitney, G. G. 1987. Some reflections on the value of old-growth forests, scientific and otherwise. *Natural Areas Journal* 7:92–99.

Whittaker, R. H., and W. A. Niering. 1965. Vegetation of the Santa Catalina Mountains, Arizona: Gradient analysis of the south slope. *Ecology* 46:429–52.

CHAPTER SEVEN

Dendroclimatology and Dendroecology in the Pinaleño Mountains

Henri D. Grissino-Mayer and Harold C. Fritts

The trees that grow on Mt. Graham and surrounding peaks have great potential for revealing factors that have shaped the environment of the Pinaleño Mountains (figure 7.1). Trees reveal information about the past history of climate and the effects of climate change upon the structure and composition of the local forest ecosystem. Because trees are excellent recorders of environmental change, biologists can use them to investigate the role of natural and anthropogenic disturbances. To "read" the environmental history recorded by the trees, we use dendrochronology (the science that analyzes the growth rings of trees) to address various questions in climate, ecology, geomorphology, and archaeology. We can thus obtain information about limiting conditions for tree growth (e.g., precipitation and temperature) in the past, which provides a unique and valuable time perspective for ecological studies (Fritts and Swetnam 1989). Possible topics for study include the frequency and areal extent of past wildfires; the frequency of insect outbreaks (e.g., western spruce budworm) and the role of these outbreaks on forest age structure and composition; and the effects of wind on tree mortality and growth. Anthropogenic disturbances that can be investigated include the effects of fire suppression on forest structure and composition, the effects of air pollutants on tree growth and mortality, the effects of possible carbon dioxide fertilization on forest productivity, and the effects of logging and road construction on landscape patterns. We can combine the results from these studies with results from other research projects, such as investigations of the ecology of the Mt. Graham red squirrel (*Tamiasciurus hudsonicus grahamensis*), to learn more about interactions between the local fauna and the environment and to help provide guidelines for sound management practices of the unique high-elevation mixed-conifer and spruce-fir forests of the Pinaleños.

Since 1988, we have urged that cross sections of trees cut during construction of the access road and telescope sites for the Mt. Graham International Observatory be collected and stored for future research projects. Although we favor the preservation of old-growth forests, we had to take advantage of this unique opportunity to learn more about the environmental history of the Pinaleños as recorded in the dendrochronological record. Obtaining cross sections from all downed trees was not feasible, though we did manage to collect cross sections and increment cores from nearly 300 trees cut during construction. Although the area represented by this collection is geographically limited, these trees nonetheless provide valuable information on the age structure of the local forests now and in the past, especially the spruce-fir forests, which have not been extensively studied in southern Arizona. We specifically targeted trees that exhibited obvious evidence of disturbance, such as the characteristic basal wound (i.e., "catface") caused by fires (figure 7.2). These specimens are now archived and stored at the University of Arizona Laboratory of Tree-Ring Research, where researchers are analyzing them in ways that reveal much about the climatic and environmental history of the Pinaleños.

Dendroclimatological Studies in the Pinaleños

Studies that investigate archaeological questions and the climate–tree growth relationship are the best known of dendrochronological applications (Schulman 1956; Fritts 1976; Dean 1978, 1988) and formed the basis of dendrochronology during its early years (Douglass 1909, 1929). One of the primary goals of dendrochronological research in the Pinaleños is to reconstruct past patterns of climate based on the average annual growth of local species. Studies have shown that different species growing in the same geographic area may respond differently to climate (Fritts et al. 1965a, 1965b; Fritts 1974; Graumlich 1989, 1993). A multispecies, dendroclimatic reconstruction for the Pinaleños should yield information on the climate history of this mountain range over and above the information obtained from just one species (cf. Fritts 1991). The potential for a multispecies dendroclimatic reconstruction is excellent given the diversity of tree species that grow within the "sky islands" of southeastern Arizona. Tree species known to exist in the Pinaleños and considered suitable for dendroclimatic studies include

Figure 7.1 This recently discovered Douglas-fir is growing on the side of a steep, rugged cliff in the vicinity of Mt. Graham. The tree has an inside ring date of A.D. 1257 and is currently the oldest known living tree in southern Arizona.

Engelmann spruce (*Picea engelmannii*); corkbark fir (*Abies lasiocarpa* var. *arizonica,* recently renamed *A. bifolia*); Rocky Mountain Douglas-fir (*Pseudotsuga menziesii* var. *glauca*); ponderosa (or Arizona) pine (*Pinus ponderosa* var. *arizonica*); southwestern white pine (*Pinus strobiformus*); and white fir (*Abies concolor*) (Johnson 1988; McLaughlin 1993).

We are only now completing the collection, laboratory preparation, and chronology development stages for the samples we collected in the

Pinaleños and are beginning a reconstruction of climate for southeastern Arizona. The reconstruction will involve several steps, of which the first involves determining to which climate variables the trees are responding using response function and correlation analyses (Fritts 1976; Buckley 1989; Fritts et al. 1991). Candidate climate variables that may be investigated include precipitation and temperature (monthly, seasonal, and annual), as well as the Palmer Drought Severity Index and

Figure 7.2 A southwestern white pine (left) and ponderosa pine (right) growing in the Emerald Peak area that exhibit characteristic basal wounds on the uphill side of the tree caused by fire. The individual ridges in the scar surface on the southwestern white pine indicate separate and repeated fire events.

Palmer Hydrological Drought Index, which integrate precipitation, temperature, soil conditions, sunlight duration, and geographic area into one index (Stahle and Cleaveland 1988; Grissino-Mayer 1988; Buckley 1989). Researchers have also used sunshine duration (Stahle et al. 1991), summer degree-days (Jacoby et al. 1985), and records of snow depth (Graumlich and Brubaker 1986) in tree-ring studies.

The second stage involves application of a transfer function that predicts climate as a function of tree growth for the entire length of the selected tree-ring chronologies (Fritts 1976, 1991). The regression between climate and the tree-ring index chronologies is usually conducted over one-half of the length of the historical observations. Researchers then use the calibration equation derived from this equation to predict climate over the remaining one-half of the observations and apply verification tests that use well-established statistics to determine how close the predicted values come to the actual climatic values for the historical period (Fritts 1991). If the calibration equation proves to be a robust estimator of climate, then researchers consider the equation verified and use it to "retrodict" climate back in time using one or more tree-ring index chronologies as the predictor variables.

Once developed, the climate reconstruction for the Pinaleños will allow us to assess the frequency, duration, and relative magnitude of past short-term and long-term climatic variations. We can specifically investigate the existence of extended periods of drought and high temperatures or above-normal rainfall and low temperatures (Stahle et al. 1988; Fritts 1991). We intend to identify any effects from a climate episode that occurred between circa A.D. 1590 and 1850, known as the "Little Ice Age" (LaMarche 1974; Luckman 1986; Briffa et al. 1990; Fritts 1991), a period that had varying effects on tree growth depending on the geographic location of the site. If the tree-ring data extend far enough back in time, we also intend to investigate the possible effects of a climate episode that occurred between circa A.D. 1000 and 1300, known as the "Medieval Warm Period" (Eckstein 1986; Briffa et al. 1990; Cook et al. 1992). Such climate studies are integral components of other dendroecological studies, such as those that investigate the relation between fire occurrence and below-normal (as well as above-normal) rainfall or those studies in which biologists need growth information from tree species known to be nonhosts for the western spruce budworm (*Choristoneura occidentalis*) (Swetnam and Lynch 1989). By integrating these results with other ongoing projects in the Pinaleños, we may determine to what extent above-normal or below-normal rainfall or temperature conditions would affect the size, distribution, and adaptive strategies of the local floral and faunal populations in the Mt. Graham area.

Given this excellent potential for dendrochronological analyses in the Pinaleños, we have delineated four additional dendroecological research projects that build upon the results obtained from the dendroclimatic analyses. These projects all have implications for the future management of the Mt. Graham red squirrel habitat and are important for any multidisciplinary studies that will be conducted in the Mt. Graham area in the future.

The History of Fire in the Pinaleños

The history of wildland fires read from tree-ring information is one of the best-known dendroecological applications in disturbance history studies (Stokes and Dieterich 1980; Swetnam 1990). Field and laboratory techniques are well established (Arno and Sneck 1977; Baisan and Swetnam 1990) and are currently being practiced and refined by many research groups within academic institutions and federal agencies (Swetnam and Dieterich 1985; Barrett and Arno 1988; Swetnam 1990; Johnson 1992).

We consider this research the primary objective in any dendroecological study within the Pinaleños because fire is an important and recurring phenomenon of any forest environment (Spurr and Barnes 1980). The biogeographic islands of southeastern Arizona, such as the Pinaleño Mountains, present a unique opportunity for the investigation of fire because the history of anthropogenic disturbances that may affect fire occurrence (e.g., grazing, logging, fuelwood cutting, and fire suppression) are well documented (Bahre 1991). Research into the fire history of the Pinaleños should specifically address several questions:

1. How often did fire occur in the mixed-conifer and spruce-fir forests in the Pinaleños prior to Euro-American settlement?
2. When did the last stand-replacing fires occur in various parts of the Pinaleños and, specifically, in known habitats for the Mt. Graham red squirrel?
3. Were these fires very large (over 400 ha), or were they confined to small areas because of topography or prevailing weather conditions?

4. What role should fire play in the future?
5. What is the current risk of fire to observatory structures and squirrel habitat?
6. What is the relation between fire occurrence in the Pinaleños and fire elsewhere in the sky islands of southern Arizona?

During preliminary field reconnaissance near the construction sites for the proposed observatory on Emerald Peak, we were surprised to discover extensive evidence of past fire (figures 7.2 and 7.3), prompting us to emphasize that resource managers should not ignore the fire history of this area in making decisions involving the newly constructed telescope complex. We found numerous logs, snags (standing dead trees), stumps, and living trees that exhibited basal wounds (fire scars), usually on the uphill side of the tree and usually very charred, indicating that fire was the most likely causal agent. All scarred surfaces contained the characteristic "ridges" caused by subsequent overgrowth by the tree after repeated fires (figure 7.2, left tree). The amount of information about fire history obtained from single specimens in the Pinaleños was exceptional by Southwestern standards. The southwestern white pine stump shown in figure 7.3 was salvaged from the access road leading to the observatory complex just prior to construction. The pith date on this tree is 1648. Between 1648 and 1880, this tree had recorded 13 major fire events—in the years 1685, 1709, 1733, 1745, 1752, 1760, 1773, 1785, 1819, 1842, 1847, 1858, and 1871. No fire had occurred after 1871, emphasizing the degree to which anthropogenic disturbances since 1880 have altered the fire regime.

At the higher elevations within and adjacent to the Mt. Graham International Observatory, fire has occurred with greater frequency than expected for these elevations (Grissino-Mayer et al. 1994). On average, a forest fire occurred about once every four years in the mixed-conifer forests below the spruce-fir forests, a frequency greater than researchers previously reported for other mixed-conifer forests in the Southwest (Ahlstrand 1980; Dieterich 1983). Of immediate concern is the fact that all fires have been suppressed in the Pinaleños for the last hundred years, first by widespread grazing after 1880, then by active fire suppression by the Forest Service beginning in the early part of the twentieth century (Bahre 1991; Grissino-Mayer et al. 1994). In the mixed-conifer forest and the mixed-conifer/spruce-fir transition forest below the telescope sites, no widespread fire has occurred since 1893 (Grissino-Mayer et al. 1994). This evidence indicates that fuels have

Figure 7.3 A southwestern white pine stump left from logging during the 1950s that exhibits a pronounced fire-scarred surface (concave area on the left side of the stump). This sample was located in the access road right-of-way to the telescope complex and was salvaged prior to road construction.

accumulated for a century in areas where fuels had periodically been reduced by fires once every four years prior to Euro-American settlement beginning around 1880. This fuel buildup and its potential for catastrophic fire should be a major concern to those responsible for managing the red squirrel habitat and the telescope complex. In stark contrast, no fire has occurred in the higher-elevation spruce-fir forest directly surrounding the telescope sites in the last three centuries, based on the age structure of the spruce and fir samples obtained. This long fire-free interval is expected given the mesic conditions of these forests, but the interval occupies the upper end of the range for spruce-fir forests, suggesting that a stand-replacing fire in the spruce-fir forests of Emerald Peak (similar to that which recently occurred in the spruce-fir forests of the nearby Chiricahua Mountains) may occur in the not-too-distant future. Widespread fires during late June and July of 1994 in southeastern Arizona further emphasize the need to learn more about the history of fire in the Pinaleños.

Air Pollution

During the last 20 years, a major focus in dendroecological research has been the possible effects of atmospheric pollution upon tree growth (Nash et al. 1975; Graybill and Rose 1989). These studies have major implications for forest management: if air pollution causes growth reduction, a corresponding decrease in timber yields will occur as well as changes in forest stand age structure and forest composition.

A study conducted in southeastern Arizona has established a possible relation between tree growth and air pollution generated by copper smelters in Pinal County (Nash et al. 1975). The trees of the Pinaleños are growing close to air pollution from the San Manuel copper smelter 64 km to the west across the Galiuro Mountains. As we show in a later section, we believe that a pollution signal may exist in the growth-ring record of trees in the Pinaleños, though it is weak and our results are still inconclusive. Establishment of a network of tree-ring chronologies for the Pinaleños would help future researchers determine if pollution has affected tree growth in southeastern Arizona and, if so, would also enable researchers to measure the magnitude and rate of any decline in growth rates.

Carbon Dioxide Fertilization

Recent dendroecological studies have shown that increases in atmospheric carbon dioxide from the burning of fossil fuels may cause an increase in the rate of photosynthesis and a subsequent increase in the amount of carbon fixed by plants (LaMarche et al. 1984; Kienast and Luxmoore 1988). This increase manifests itself in the growth-ring record as a gradual increase in the widths of annual rings. Currently, researchers are debating this issue because conflicting results have been reported from geographically homogeneous areas (Graumlich 1991;

Graybill and Idso 1993). Trees growing in the Pinaleños could help resolve this controversy. Studies that investigate effects of carbon dioxide enrichment upon tree growth usually are conducted in high-elevation sites because trees growing at high elevations appear to be most susceptible to any change in the atmospheric content of carbon dioxide (Graybill 1987; Graumlich 1991). Tree-ring chronologies in the vicinity of the observatory can thus provide baseline information for long-term examination of the carbon dioxide effect in southern Arizona.

Cone Crops, Tree Rings, and the Red Squirrel

Researchers also can use tree rings to study the relationship between cone crop production and radial growth of trees (Eis et al. 1965; Chalupka et al. 1976; Martyanov and Batalov 1990). During years in which trees produce cone and seed crops, researchers observe a decrease in ring widths because a large part of the carbon allocation is shifted to reproduction rather than radial and vertical growth. Such studies are possible because tree growth can be modeled as an aggregate of physiological age, climate, and endogenous (originating from within the stand) and exogenous (originating from outside the stand) factors (Cook 1987). Effects upon tree growth arising from exogenous factors (e.g., fire, insect damage) can be controlled for during the sampling process whereas effects due to natural aging are minimized during the standardization process. Once we have established the tree growth–climate relationship, we can account for climate effects as well. This process of elimination leaves us with a residual tree-ring series that may contain a record of cone crop production. Thus, dendrochronological analysis can identify past years in which the trees of the Pinaleños have produced significant cone crops and can indicate the relative frequency with which significant cone crops occur in the subalpine and mixed-conifer forests of the Pinaleños. Such an analysis should be of particular importance for managers of the Pinaleño forests, who are currently very concerned about the relation between cone crops and the population dynamics and future survival of the Mt. Graham red squirrel population (USFS 1988).

Modeling Environmental Conditions in the Pinaleños

A recently developed empirical model called PRECON (for factors "preconditioning" tree growth) can identify certain environmental conditions that have influenced past growth or can simulate growth condi-

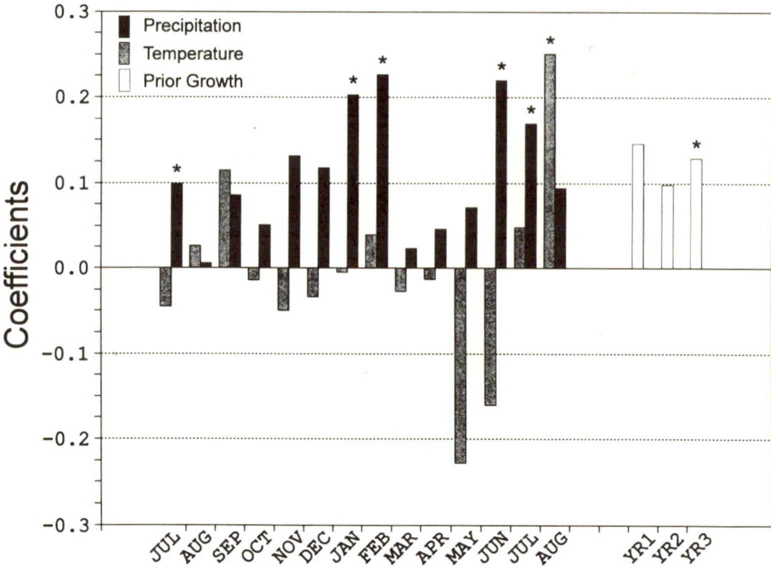

Figure 7.4 Response function coefficients for monthly precipitation and temperature (1896–1983) for Douglas-fir trees growing near Mt. Graham. Precipitation during the previous winter and current summer and temperature late in the growing season have significant, positive effects upon tree growth. * indicates a significant ($p < 0.05$) coefficient.

tions that may result from future environmental changes (Fritts et al. 1991; Fritts and Shashkin 1994). This model uses multivariate techniques (Fritts 1962) to calibrate certain climate variables, such as monthly temperature, precipitation, or Palmer Drought Severity Indices, with certain standardized tree-ring measurement series, such as total ring width, earlywood width, latewood width, or maximum latewood density (Fritts et al. 1991). Researchers can apply this model to tree-ring chronologies developed in the Pinaleños to determine which climate variables are most significant for the growth of the trees; this information is fundamental to most of the problems discussed previously.

To illustrate, we calculated coefficients for a bootstrapped response function over the calibration period 1896–1983 between climate and tree growth of two different species: (1) Douglas-fir trees growing on a south-facing slope at an elevation of 2,900 m about 1 km west of the entrance to the access road to the observatory on Emerald Peak and (2) corkbark fir trees that had grown at the site of the astrophysical complex on Emerald Peak at an elevation of 3,200 m (Grissino-Mayer and

Swetnam 1992). We regressed tree growth on orthogonal transformations of 14 monthly precipitation and 14 monthly temperature variables beginning with July of the previous year and ending with August of the current year. We included prior growth from the three previous years in these analyses because a preconditioning effect of physiological activity occurs during the previous growing seasons that becomes translated to the current year's tree growth (Fritts et al. 1965b; Fritts 1976).

For Douglas-fir, we found significant positive coefficients for precipitation during July of the previous year and during January, February, June, and July of the current year over the period 1896–1983 (figure 7.4). This positive relationship is expected because Douglas-firs respond mainly to precipitation in the preceding autumn and winter (as snowfall), as well as precipitation during the summer monsoon. The coefficients for precipitation in November and December were also high and positive. For corkbark fir, we found significant coefficients for precipitation during November and December of the previous year and during February, March, and June of the current year (figure 7.5). The

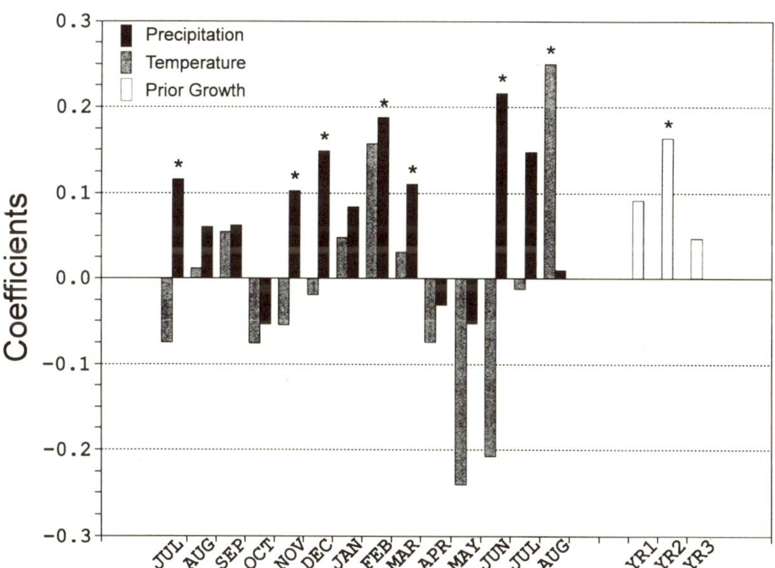

Figure 7.5 Response function coefficients for monthly precipitation and temperature (1896–1983) for corkbark fir trees growing on Mt. Graham. This species has a longer response to winter precipitation than does Douglas-fir.

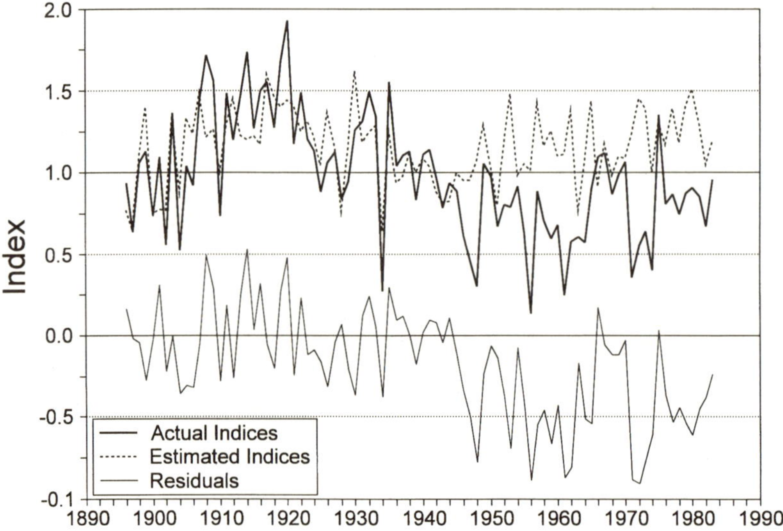

Figure 7.6 Actual and estimated indices of tree growth and their residuals developed for ponderosa pine trees growing on Green Mountain, obtained by calibration with climate from 1896 to 1940. The decline of actual tree growth beginning around 1945 suggests that some external factor has adversely affected tree growth in the Santa Catalina Mountains.

higher-elevation corkbark fir appears to respond to winter precipitation over a longer duration (November through February) than does the Douglas-fir. This longer response supports our suggestion that by using different species researchers can obtain better information about climate–tree growth relationships. Growth of both Douglas-fir and corkbark fir is directly related to temperature in August at the end of the growing season, perhaps because temperature during this month prolongs growth activities.

Because we are able to model the climate–tree growth relationship with PRECON, we can use this information to investigate possible changes in tree growth that may be associated with increasing atmospheric pollutants. For this analysis, we chose a ponderosa pine tree-ring index chronology developed from trees growing at the Green Mountain site near the observatory in the Santa Catalina Mountains just north of Tucson, Arizona, and adjacent to the large San Manuel smelter in the San Pedro Valley to the east. Using PRECON, we calibrated the chronology with climate for the period 1896–1940 and used the statistical relationship from the calibration to calculate the growth that

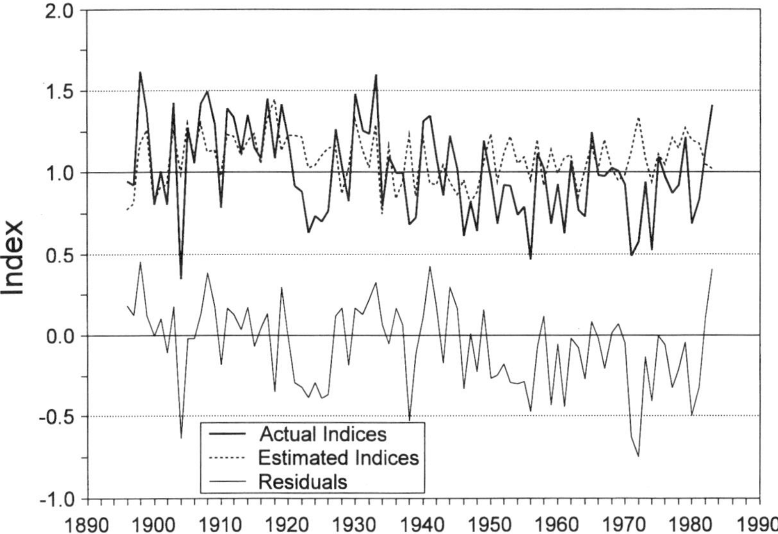

Figure 7.7 Actual and estimated indices of tree growth and their residuals developed for Douglas-fir trees growing near Mt. Graham, obtained by calibration with climate from 1896 to 1940. A reduction in tree growth similar to but less pronounced than that of the ponderosa pine has occurred beginning around 1945.

would have been expected only from climate variations over the period 1941 to 1983. The curious growth decline in the residuals beginning around 1945 (figure 7.6) clearly indicates that some factor, perhaps smelter emissions, has reduced ponderosa pine growth in the Santa Catalina Mountains below that expected from observed climate variations.

We conducted a similar analysis using the Douglas-fir tree-ring chronology developed for the Pinaleños (Grissino-Mayer and Swetnam 1992). Like that of the Green Mountain ponderosa pines, the growth of these Douglas-fir trees is below that expected due only to normal climate variations based on the 1896–1940 calibration (figure 7.7), although the difference is less than that for the Green Mountain ponderosa pines. The timing of this reduction is similar to the onset of growth reduction observed in Green Mountain ponderosa pine, suggesting a possible regional explanation for the growth reductions. Until we can develop a direct, mechanistic link, we can only hypothesize that these growth reductions are caused by air pollutants from local smelters. Other factors not yet considered could also affect tree growth on a regional level.

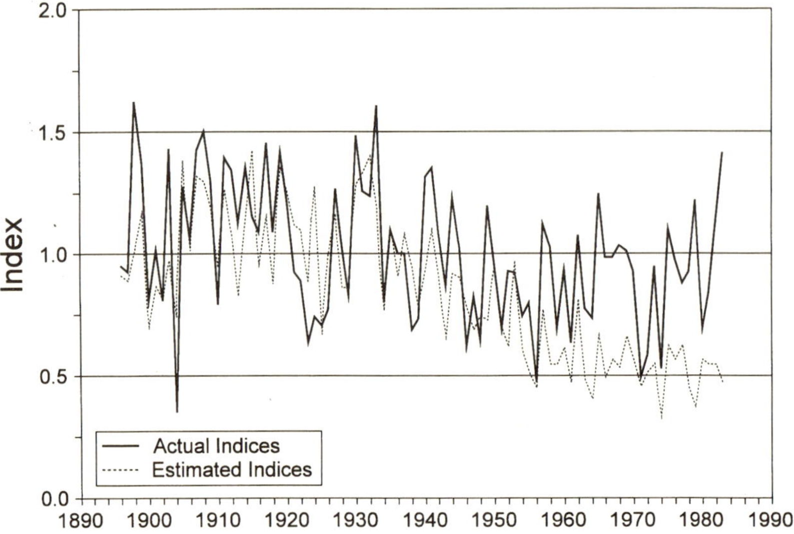

Figure 7.8 Results of a simulation in which precipitation is reduced by 0.04% annually beginning in 1940, showing a marked reduction in tree growth of Douglas-fir trees growing near Mt. Graham.

We would also need to show that these growth reductions are occurring in most forests of southeastern Arizona. Of 19 tree-ring chronologies developed for southeastern Arizona contained in the International Tree-Ring Data Bank, only 7 showed some degree of growth reduction below that expected due to climate after 1940. Several reasons may explain why some trees within the region show a reduction in tree growth while others do not. First, the topography of southeastern Arizona is basin and range, with mountain ranges trending north and south. This topography may cause differential effects upon tree growth by pollutants depending upon whether the trees are located downwind or upwind from the pollution source. Second, the compass-direction aspect of the location of the trees sampled may ameliorate pollution effects due to differences in microclimatic characteristics. Third, tree-ring chronologies developed from low-elevation trees may show greater effects than trees at higher elevations because emissions are more concentrated in the lower elevations due to inversions common in the valleys of southeastern Arizona. Finally, different species may have different responses to pollution just as they have different responses to climate.

Using the climate–tree growth relationship, the model can also simulate the probable effects of climate change upon tree growth. Using PRECON, we entered a 0.04% constant reduction in annual (January–December) precipitation beginning in the year 1940 to observe the effects this reduction would have upon trees in the Pinaleños. After 1940, the declining levels of precipitation would, as we expected, reduce tree growth (figure 7.8), but the effects under real-world conditions would perhaps be more moderate because of other factors that affect tree growth. We then simulated the expected changes in tree growth that would result from an increase in summer temperature. Again as we expected, the simulated increase of about two degrees in annual temperature would increase the growth of trees (figure 7.9), a pattern that has implications for investigating the effects upon tree growth by the hypothetical increase in temperatures attributed by many climatologists to a worldwide "greenhouse effect." Studies of past climate must therefore consider that tree growth is not just dependent upon one type of climate variable, such as precipitation, but that the climate variables themselves are interconnected and inseparable. A change in one climate variable may actually be masked by a subsequent change in the opposite direction by another. This analysis is only a

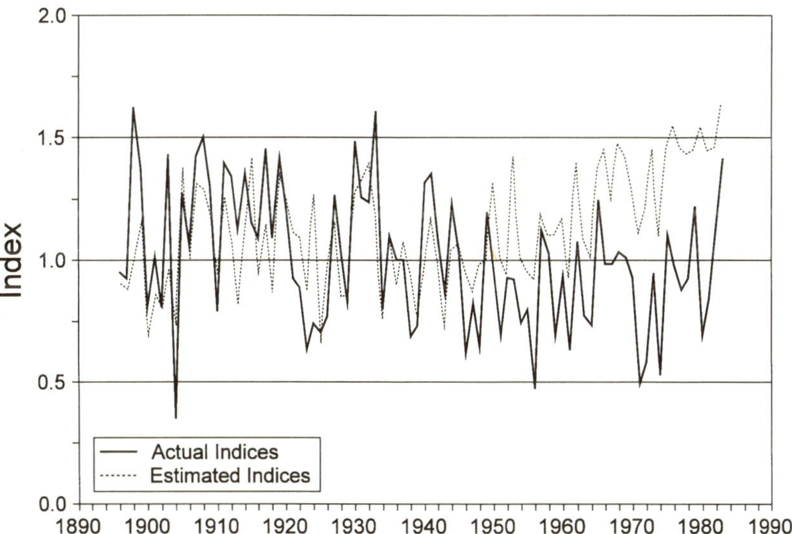

Figure 7.9 Results of a simulation in which summer temperature is increased by 0.1% annually beginning in 1940, showing a subsequent increase in growth rates by Douglas-fir trees growing near Mt. Graham.

beginning because it involves only two of the many microclimatically varying sites available in the Pinaleños. However, our preliminary analyses suggest that the dominant tree species in the Pinaleños, and in southeastern Arizona generally, may be responding to the preliminary effects of air pollution and climatic change, though investigators need much more evidence before accepting such conclusions.

Epilogue

In 1991, we discovered stands of ancient Douglas-fir and southwestern white pine trees growing just to the west of the access road leading up to the observatory (figure 7.1). These trees have since provided us with a continuous tree-ring chronology dating back to A.D. 1106, over 300 years longer than any tree-ring chronology that had been developed in southern Arizona. This chronology will, in turn, yield the longest climate reconstruction yet developed for this area. This discovery emphasizes the immense potential the trees of Mt. Graham have for providing information on the past and present environment of the various habitats in the "sky islands" of southeastern Arizona. These ancient trees are in no danger of destruction due to road building or observatory construction and exist in what is considered a very protective and "fireproof" habitat, growing directly on granite cliffs. Because commercial logging has been discontinued in the Pinaleños by the Forest Service (USFS 1988), these trees are ensured continued protection. During any future construction, dendrochronologists will be consulted to ensure that cross sections from all trees cut for observatory construction will be available for study.

Acknowledgments

Funding was provided by the University of Arizona through the Mt. Graham Red Squirrel Study Committee, comprising representatives of federal and state agencies. We thank the Steward Observatory and the Forest Service for help in the collection of material. We are grateful to the following individuals for help in the field or laboratory: Rex Adams, Didier Bert, Kevin Bonine, Peter Brown, Jean-Luc Dupouey, Mike Farrell, Robert Heckman, Emily Heyerdahl, Esther Jansma, Michéle Kaennel, John Maingi, Ronald Martin, Jim Miller, Kiyomi Morino, Martin Munro, James Riser, Matthew Riser, Robert Shay, Paul Sheppard, Denise Slama, James Speer, and Tom Swetnam. Tree-ring data discussed or analyzed were obtained from the International Tree-Ring Data Bank. Brendan Buckley kindly supplied two additional Engelmann spruce chronologies collected near Mt. Graham.

References

Ahlstrand, G. M. 1980. Fire history of a mixed conifer forest in Guadalupe Mountains National Park. In *Proceedings of the Fire History Workshop, October 20–24, 1980, Tucson, Arizona,* ed. M. A. Stokes and J. H. Dieterich, pp. 4–7. General Technical Report RM-81. U.S. Department of Agriculture, Forest Service, Ft. Collins, Colo.

Arno, S. F., and K. M. Sneck. 1977. *A Method for Determining Fire History in Coniferous Forests of the Mountain West.* General Technical Report INT-42. U.S. Department of Agriculture, Forest Service, Ogden, Utah.

Bahre, C. J. 1991. *A Legacy of Change: Historic Human Impact on Vegetation of the Arizona Borderlands.* University of Arizona Press, Tucson.

Baisan, C. H., and T. W. Swetnam. 1990. Fire history on a desert mountain range: Rincon Mountain Wilderness, Arizona, U.S.A. *Canadian Journal of Forest Research* 20:1559–69.

Barrett, S. W., and S. F. Arno. 1988. *Increment-Borer Methods for Determining Fire History in Coniferous Forests.* General Technical Report INT-244. U.S. Department of Agriculture, Forest Service, Ogden, Utah.

Briffa, K. R., T. S. Bartholin, D. Eckstein, P. D. Jones, W. Karlén, F. H. Schweingruber, and P. Zetterberg. 1990. A 1,400-year tree-ring record of summer temperatures in Fennoscandia. *Nature* 346:434–39.

Buckley, B. M. 1989. Dendroclimatic aspects of Engelmann spruce in southern Arizona and New Mexico. M.S. thesis, Arizona State University, Tempe.

Chalupka, W., M. Giertych, and Z. Krolikowski. 1976. The effect of cone crops in Scots pine on tree diameter increment. *Arboretum Kornickie* 21:361–66.

Cook, E. R. 1987. The decomposition of tree-ring series for environmental studies. *Tree-Ring Bulletin* 47:37–59.

Cook, E., T. Bird, M. Peterson, M. Barbetti, B. Buckley, R. D'Arrigo, and R. Francey. 1992. Climatic change over the last millennium in Tasmania reconstructed from tree-rings. *The Holocene* 2:205–17.

Dean, J. S. 1978. Tree-ring dating in archaeology. In *Miscellaneous Collected Papers 19-24,* ed. J. D. Jennings, pp. 129–63. University of Utah Anthropological Papers 99. Salt Lake City.

Dean, J. S. 1988. Dendrochronology and paleoenvironmental reconstruction on the Colorado Plateaus. In *The Anasazi in a Changing Environment,* ed. G. J. Gumerman, pp. 119–67. Cambridge University Press, New York.

Dieterich, J. H. 1983. Fire history of southwestern mixed conifer: A case study. *Forest Ecology and Management* 6:13–31.

Douglass, A. E. 1909. Weather cycles in the growth of big trees. *Monthly Weather Review* 37:225–37.

Douglass, A. E. 1929. The secret of the Southwest solved by talkative tree rings. *National Geographic Magazine* 56:736–70.

Eckstein, D. 1986. Temperature fluctuations in western Europe during the last 1000 years as derived from tree rings and other proxy indicators. In *Regional Resource Management 1*, ed. L. Kairiukstis, pp. 33–45. International Institute for Applied Systems Analysis, Laxenburg, Austria.

Eis, S., E. H. Garman, and L. F. Ebell. 1965. Relation between cone production and diameter increment of Douglas-fir (*Pseudotsuga menziesii* [Mirb.] Franco), grand fir (*Abies grandis* [Dougl.] Lindl.), and western white pine (*Pinus monticola* Dougl.). *Canadian Journal of Botany* 43:1553–59.

Fritts, H. C. 1962. An approach to dendroclimatology: Screening by means of multiple regression techniques. *Journal of Geophysical Research* 67:1413–20.

Fritts, H. C. 1974. Relationships of ring widths in arid-site conifers to variations in monthly temperature and precipitation. *Ecological Monographs* 44:411–40.

Fritts, H. C. 1976. *Tree Rings and Climate*. Academic Press, New York.

Fritts, H. C. 1991. *Reconstructing Large-Scale Climatic Patterns from Tree-Ring Data*. University of Arizona Press, Tucson.

Fritts, H. C., and A. V. Shashkin. 1994. Modeling tree-ring structure as related to temperature, precipitation, and day length. In *Tree Rings as Indicators of Ecosystem Health*, ed. T. E. Lewis. Lewis Publishers, Boca Raton, Fla.

Fritts, H. C., D. G. Smith, J. W. Cardis, and C. A. Budelsky. 1965a. Tree-ring characteristics along a vegetational gradient in northern Arizona. *Ecology* 46:393–401.

Fritts, H. C., D. G. Smith, and M. A. Stokes. 1965b. The biological model for paleoclimatic interpretation of Mesa Verde tree-ring series. *American Antiquity* 31:101–21.

Fritts, H. C., and T. W. Swetnam. 1989. Dendroecology: A tool for evaluating variations in past and present forest environments. *Advances in Ecological Research* 19:111–88.

Fritts, H. C., E. A. Vaganov, I. V. Sviderskaya, and A. V. Shashkin. 1991. Climatic variations and tree-ring structure in conifers: Empirical and mechanistic models of tree-ring width, number of cells, cell size, cell-wall thickness and wood density. *Climate Research* 1:97–116.

Graumlich, L. J. 1989. The utility of long-term records of tree growth for improving forest stand simulation models. In *Natural Areas Facing Climate Change*, ed. G. P. Malanson, pp. 39–49. SBP Academic Publishing, The Hague, Netherlands.

Graumlich, L. J. 1991. Subalpine tree growth, climate, and increasing CO_2: An assessment of recent growth trends. *Ecology* 72:1–11.

Graumlich, L. J. 1993. A 1000-year record of temperature and precipitation in the Sierra Nevada. *Quaternary Research* 39:249–55.

Graumlich, L. J., and L. B. Brubaker. 1986. Reconstruction of annual temperature (1590-1979) for Longmire, Washington, derived from tree rings. *Quaternary Research* 25:223–34.

Graybill, D. A. 1987. A network of high elevation conifers in the western U.S. for detection of tree-ring growth response to increasing atmospheric carbon dioxide. In *Proceedings of the International Symposium on Ecological Aspects of Tree-Ring Analysis,* ed. G. C. Jacoby Jr. and J, W. Hornbeck, pp. 464–74. Publication CONF-8608144. U.S. Department of Energy, Washington, D.C.

Graybill, D. A., and S. B. Idso. 1993. Detecting the aerial fertilization effect of atmospheric CO_2 enrichment in tree-ring chronologies. *Global Biogeochemical Cycles* 7:81–95.

Graybill, D. A., and M. R. Rose. 1989. Analysis of growth trends and variation in conifers from Arizona and New Mexico. In *Effects of Air Pollution on Western Forests,* ed. R. K. Olson and A. S. Lefohn, pp. 393–407. Air and Waste Management Association, Pittsburgh.

Grissino-Mayer, H. D. 1988. Tree rings of shortleaf pine (*Pinus echinata* Mill.) as indicators of past climatic variability in north central Georgia. M.A. thesis, University of Georgia, Athens.

Grissino-Mayer, H. D., C. H. Baisan, and T. W. Swetnam. 1994. Fire history and age structure analyses in the mixed-conifer and spruce-fir forests of Mount Graham, southeastern Arizona. Mt. Graham Red Squirrel Study Committee, U.S. Fish and Wildlife Service, Phoenix, Ariz. (Unpublished.)

Grissino-Mayer, H. D., and T. W. Swetnam. 1992. Dendroecological research on Mt. Graham: Development of tree-ring chronologies for the Pinaleño Mountains. Mt. Graham Red Squirrel Study Committee, U.S. Fish and Wildlife Service, Phoenix, Ariz. (Unpublished.)

Jacoby, G. C., Jr., E. R. Cook, and L. D. Ulan. 1985. Reconstructed summer degree days in central Alaska and northwestern Canada since 1524. *Quaternary Research* 23:18–26.

Johnson, E. A. 1992. *Fire and Vegetation Dynamics: Studies from the North American Boreal Forest.* Cambridge University Press, New York.

Johnson, W. T. 1988. Flora of the Pinaleño Mountains, Graham County, Arizona. *Desert Plants* 8:147–62, 175–91.

Kienast, F., and R. J. Luxmoore. 1988. Tree-ring analysis and conifer growth responses to increased atmospheric CO_2 levels. *Oecologia* 76:487–95.

LaMarche, V. C., Jr. 1974. Paleoclimatic inferences from long tree-ring records. *Science* 183:1043–48.

LaMarche, V. C., Jr., D. A. Graybill, H. C. Fritts, and M. R. Rose. 1984. Increasing atmospheric carbon dioxide: Tree ring evidence for growth enhancement in natural vegetation. *Science* 225:1019–21.

Luckman, B. H. 1986. Reconstruction of Little Ice Age events in the Canadian Rocky Mountains. *Geographie physique et Quaternaire* 40:17–28.

McLaughlin, S. P. 1993. Additions to the flora of the Pinaleño Mountains, Arizona. *Journal of the Arizona-Nevada Academy of Science* 27:5–32.

Martyanov, N. A., and A. A. Batalov. 1990. Relationships between radial increment of conifers and intensity of cone production. *Lesovedenie* (0)2:30–36.

Nash, T. H., III, H. C. Fritts, and M. A. Stokes. 1975. A technique for examining nonclimatic variation in widths of annual tree rings with special reference to air pollution. *Tree-Ring Bulletin* 35:15–24.

Schulman, E. 1956. *Dendroclimatic Changes in Semiarid America.* University of Arizona Press, Tucson.

Spurr, S. H., and B. V. Barnes. 1980. *Forest Ecology.* John Wiley & Sons, New York.

Stahle, D. W., and M. K. Cleaveland. 1988. Texas drought history reconstructed and analyzed from 1698 to 1980. *Journal of Climate* 1:59–74.

Stahle, D. W., M. K. Cleaveland, and R. S. Cerveny. 1991. Tree-ring reconstructed sunshine duration over central USA. *International Journal of Climatology* 11:285–95.

Stahle, D. W., M. K. Cleaveland, and J. G. Hehr. 1988. North Carolina climate changes reconstructed from tree rings: A.D. 372 to 1985. *Science* 240: 1517–19.

Stokes, M. A., and J. H. Dieterich, eds. 1980. *Proceedings of the Fire History Workshop.* General Technical Report RM-81. U.S. Department of Agriculture, Forest Service, Ft. Collins, Colo.

Swetnam, T. W. 1990. Fire history and climate in the southwestern United States. In *Effects of Fire Management of Southwestern Natural Resources,* ed. J. S. Krammes, pp. 6–17. General Technical Report RM-191. U.S. Department of Agriculture, Forest Service, Ft. Collins, Colo.

Swetnam, T. W., and J. H. Dieterich. 1985. Fire history of ponderosa pine forests in the Gila Wilderness, New Mexico. In *Proceedings—Symposium and Workshop on Wilderness Fire,* ed. J. E. Lotan, B. M. Kilgore, W. C. Fischer, and R. W. Mutch, pp. 390–97. General Technical Report INT-182. U.S. Department of Agriculture, Forest Service, Ogden, Utah.

Swetnam, T. W., and A. M. Lynch. 1989. A tree-ring reconstruction of western spruce budworm history in the southern Rocky Mountains. *Forest Science* 35:962–86.

USFS. 1988. *Final Environmental Impact Statement, Proposed Mt. Graham Astrophysical Area, Pinaleño Mountains, Coronado National Forest.* U.S. Department of Agriculture, Forest Service, Southwestern Region, Tucson, Ariz.

The Biogeography of the Pinaleños

Past and Present

Conservation biology is the subdiscipline of ecology that attempts to develop sound strategies for coping with the crucial problem of the preservation and restoration of organic diversity. One job of conservation biologists is to attempt to understand the changes that occur in the structure of [ecological] communities . . . as those communities become more and more simplified and fragmented.

PAUL R. EHRLICH, *The Machinery of Nature*

The three chapters of Part 4 define and explore the biogeographical context necessary for an understanding of the origin and maintenance of biological diversity and ecological structure in the terrestrial island archipelago that includes the Pinaleño Mountains, with Mt. Graham and the observatory at Emerald Peak. The exploration of such a broadly conceptual context provides a background, a fundamental level of scientific understanding, essential for all future planning for biological conservation in the Pinaleños and its entire inland archipelago of mountain islands.

In the first chapter, Russell Davis creates the archipelago context, setting the stage for many future investigations and several ensuing chapters. He develops ideas and inferences from a comparison with the extraordinary marine Malay Archipelago. Next, Frederick Gehlbach examines in depth the structure of bird communities across the archipelago and finds some remarkable regularities. Bruce Patterson then investigates the patterns of Holocene extinction among small mammal species throughout the archipelago and how these extinctions have shaped present-day mammalian associations in the Southwest.

These essays clearly reveal that the Pinaleños cannot be considered in isolation from the whole archipelago, either in space or in time. These large scales of great biological interest are not particularly familiar to those responsible for land and wildlife management. These scales are central to future understanding of a biological diversity shaped by forces of extinction, colonization, and demographic replacement in situ, as well as by human activities. Thus, as we found daunting complexity in the political and social spheres explored in Part One, complexity on the biological side seems somehow similar in scale and importance, though the two spheres involve questions and traditions worlds apart.

Though not wanting in imperfections and uncertainties—as with all science—these chapters and all subsequent ones provide examples of the struggle for rigor in ecological and population analysis, especially in long-term studies. Such rigor must become the hallmark of future research intended to provide effective solutions to problems of conservation biology.

CHAPTER EIGHT

The Pinaleños as an Island in a Montane Archipelago

Russell Davis

Islands have been of great importance in the development of evolutionary, ecological, and biogeographical theory. Carolus Linnaeus, the great Swedish botanist, visualized a tropical island with a high mountain as the logical "Garden of Eden" on which plants and animals were created and from which they eventually dispersed to Europe. Later, the theory of evolution by natural selection, co-discovered by Alfred Russel Wallace and Charles Darwin, was to a considerable extent shaped by Wallace's biogeographical work on the islands of the Malay Archipelago and Darwin's experiences on the Galapagos Archipelago and other islands. Considerably later, Robert H. MacArthur and Edward O. Wilson (1967) studied the diversity of island species assemblages controlled by extinction and immigration, thereby combining ecology and classical biogeography. Of course, to suggest that the beginning of an understanding of natural selection and species diversity would not have happened if the seas had lacked islands would surely be extreme; nevertheless, observations on islands have certainly led to the development of central concepts in ecology and evolutionary biology, and these concepts might have been slower to come without the insular examples and models based upon them.

The primary scientific value of islands comes from their simplicity. Compared to a similar-sized mainland area, an island has discrete boundaries, fewer species, and smaller population sizes, making ecological and evolutionary hypotheses easier to formulate and test. A study that is impossible on a mainland may be quite tractable and well-controlled on a small island.

Mountain Islands

Large marine islands, both oceanic and continental, received most of the early attention from biologists. However, biologists soon realized that since the ecological and evolutionary significance of an island comes from its isolated habitats, an ecological island need not be large or even surrounded by water. We now understand that ecological islands come in a wide variety of sizes and types. Examples of ecological islands include tiny antibiotic spots in the middle of a bacterial culture plate, clumps of thistle plants and their associated insects, riparian situations surrounded by the desert, "cow pies" in the middle of a pasture, small islands a few meters across in the St. Lawrence River, and floating clumps of vegetation; all of these contain isolated habitats and each has been a setting for ecological research. The same idea—that an ecological island consists simply of an isolated habitat or set of habitats—has been extended to include isolated mountains. Unquestionably, for plants restricted to the conditions of temperature and moisture at the top of an isolated "sky island" and for the animals restricted to the habitats defined by these plants, the unsuitable habitat at the base of the mountain is ecologically a "sea." Similarly, the "mainland" for a well-isolated mountain island must surely be some nearby large mountain range (with the same habitat or habitats) that was originally the source of species. Each isolated montane island thus contains a subset of all the species (extant or ancestral) that occur on the mainland mountain range.

It is also obvious that in a terrestrial archipelago the size of the "islands" and the extent and composition of the "sea" (and thus the degree of isolation of the islands) depend in large part on how the islands are defined ecologically. Thus, islands that by definition consist of both the forest and the woodland portions of isolated mountains are few and generally large, and they are surrounded by a sea consisting of "coastal" semidesert grasslands and "deep-water" desertscrub. In contrast, islands that are defined as consisting of only the forest portion of isolated mountains (i.e., ponderosa pine, mixed-conifer forest, spruce-fir forest, and other high-elevation habitats) are more numerous and generally much smaller, and in this case the surrounding sea consists of coastal woodlands (oak, pine-oak, and pinyon-juniper) and deep-water semidesert grasslands and desertscrub (figure 8.1).

Marine islands are of two types: continental islands (those that are on the continental shelf and become connected with the mainland

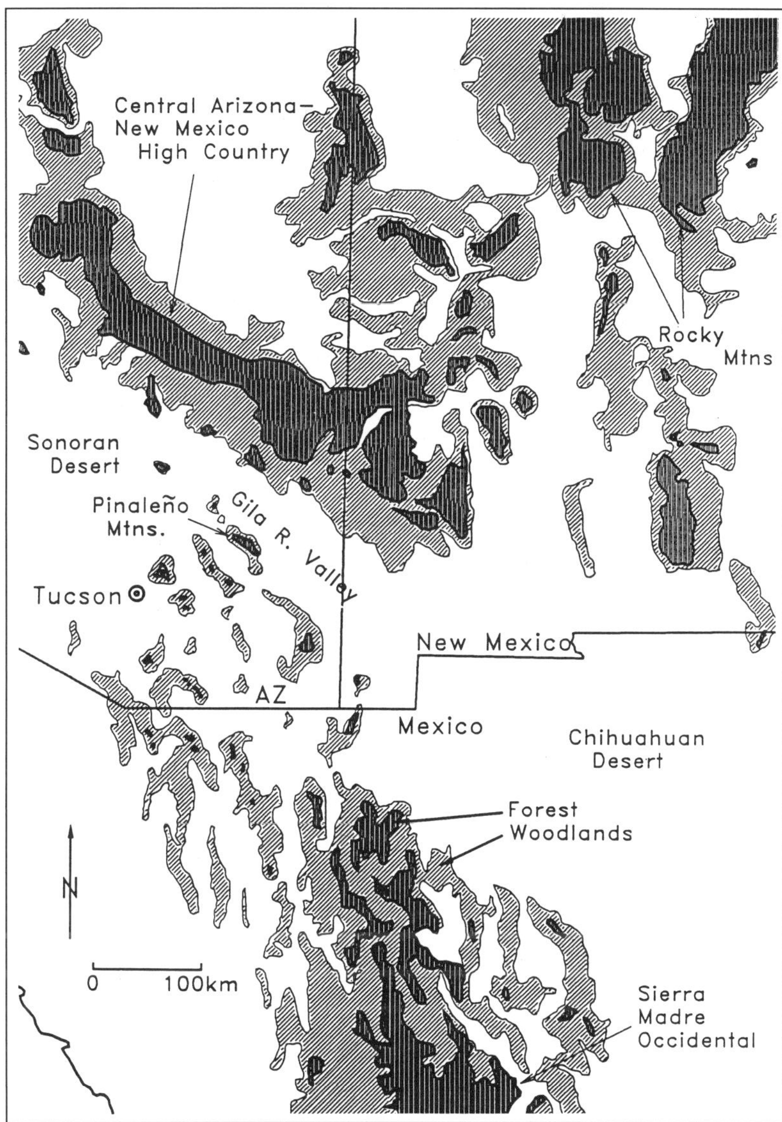

Figure 8.1 The terrestrial island archipelago in New Mexico, Arizona, and northern Mexico. Forest islands (ponderosa, mixed-conifer, and spruce-fir forests, and other high elevation habitats) are shown in dark, vertical stippling; woodland islands (oak, pine-oak, and pinyon-juniper) are shown in lighter, diagonal stippling. Because of the scale used, several small forest islands in southeastern Arizona and northern Mexico were omitted. (Modified from Brown and Lowe 1983.)

when sea levels drop), and oceanic islands (those that were never connected to any mainland). Similar distinctions occur among montane islands; in this case, "continental islands" are isolated mountains whose habitats were once connected to the mainland mountain range habitats, whereas "oceanic islands" are isolated mountains whose habitats were never connected to mainland habitats. James H. Brown (1971, 1978) was the first to successfully test the equilibrium theory of MacArthur and Wilson (1967) by applying concepts developed on marine islands to a terrestrial mountain island system. He was able to do this because he selected an island system with adequate species presence and absence data (the archipelago of montane islands in the Great Basin), because he used an appropriate ecological definition of the islands of concern (woodlands and forests), and because he was able to determine the mainlands for each island in this archipelago (the Rocky Mountains to the east and the Sierra Nevada to the west). In consequence, his analysis allowed him to conclude that the mammalian fauna on these montane islands was in nonequilibrium and that all the islands in the system were historically of the continental type.

The Montane Island System of the Southwestern United States and Northwestern Mexico

The montane island system that occurs in the eastern half of Arizona, northern Mexico, and New Mexico is one that has been relatively little studied. This terrestrial archipelago (figure 8.1) consists of a major chain that extends southward from the southwestern end of the southern extensions of the Rocky Mountains in northern New Mexico, southward through the central Arizona–New Mexico high country, through a series of small but important mountains in southeastern Arizona and northern Sonora, finally reaching the northern end of the Sierra Madre Occidental in Chihuahua, Mexico. A minor chain also extends southward from the southern extensions of the Rocky Mountains in eastern New Mexico; this chain is situated on the eastern side of the Rio Grande River and ends at the New Mexico–Texas border.

Although differing in details, several studies of biogeographical provincialism have demonstrated the uniqueness of the species assemblages of various parts of this terrestrial archipelago in Arizona and New Mexico (e.g., Hagmeier and Stults 1964; McLaughlin 1989). Of special significance, however, is the study by James S. Findley and William Caire (1977), which expanded the study of mammalian provincialism in the southwestern United States begun by Hagmeier and Stults (1964) and extended this study southward into Mexico. In addition to increasing the number of recognizable provinces for this region, Findley and Caire demonstrated the extension of one of these into the northern portions of Sonora and Chihuahua, Mexico, and established the distinctiveness of another new province that included the Sierra Madre Occidental. Findley and Caire made it clear that the isolated mountains of northern Mexico and also the Sierra Madre Occidental are biologically akin to the borderlands mountains of Arizona and New Mexico.

Moreover, by comparing the total area and species richness (number of species) of the Sierra Madre Occidental with those of the more northern mountain islands of the region (figure 8.1), one can see that the nearest match to the Sierra Madre Occidental in size and in diversity (but not in species composition) is the Rocky Mountain cordillera far to the north at the opposite end of the archipelago. At least among mammals, all the species occurring on the montane islands in the archipelago (with the exception of a few endemics) can be found on one or the other of these two major biogeographical regions. Thus, the two mainlands for this archipelago have been located—one in the south and one in the north.

A Biogeographical Analogy

We now understand that we are dealing here with a system of mountain islands that forms an archipelago appropriately located midway between two distant mainlands. I suggest that insight into the processes that produced and continue to control the assemblages of species and the species richness within this terrestrial archipelago can be gained best by comparing the traits that this system shares with another archipelago—one that is undoubtedly the most complex of them all. As unlikely as the comparison might seem at first glance, this terrestrial archipelago of mountain islands shows several characteristics that seem strikingly similar to those of the marine islands in Wallace's much larger Malay Archipelago (figure 8.2; cf. figure 8.1). In this biogeographical analogy, if the montane archipelago of mountain islands in eastern Arizona, northwestern New Mexico, and northern Mexico corresponds to the Malay Archipelago, then the Sierra Madre Occidental in the south (in Mexico) must correspond to the Australian mainland, and the Rocky Mountains to the north must correspond to the Asian mainland.

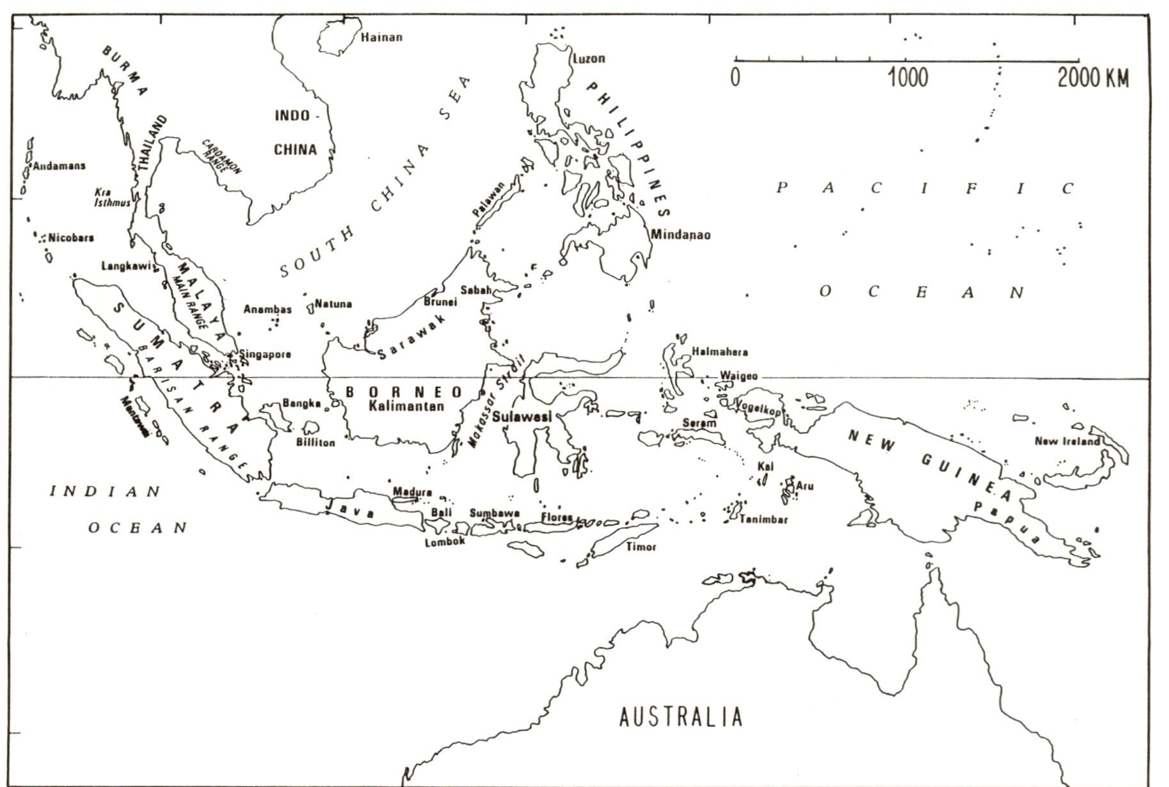

Figure 8.2 The marine island archipelago in southeastern Asia. (Modified from van Balgooy 1987.)

As is always the case, the extent of the "sea" in the montane archipelago depends upon the criteria used to define an island. It may consist of desertscrub, semidesert grasslands, and even woodlands; but in every case these are lower-elevation habitats that are unsuitable for the mountain-dwelling flora and fauna living on the "islands" and the "mainlands." This sea, in one form or another, is continuous between all the montane islands and between the islands and the mainlands, just as it is in the marine archipelago. In the terrestrial archipelago, the sea extends to the west as the Sonoran Desert and to the east as the Chihuahuan Desert. The comparable geographical features in the region of southeastern Asia are the Indian Ocean to the west and the South Pacific Ocean to the east.

In this biogeographical analogy, the central Arizona–western New Mexico high country, along with the Mogollon Plateau, with its large size and proximity to the northern mainland, would best correspond to Borneo (plus perhaps also Sumatra and Java) of the Malay Archipelago. This high country has a rich flora and fauna that is most similar to that of the northern mainland (the Rocky Mountains), just as Borneo shares a great similarity with the Asian mainland. The similarity that Borneo, Sumatra, and Java share with each other and with the Asian mainland is explained by the fact that these three large islands lie on the Asian continental shelf (i.e., they are continental islands); they were interconnected to each other and to the Asian mainland during the Pleistocene (most recent series of ice ages) when sea levels were lowered because a significant portion of the world's water was tied up in glaciers (Heaney and Rickart 1990).

During Pleistocene glacial periods, wetter-colder climates in southwestern North America caused present-day high-elevation habitats to occur at lower elevations. In consequence, the montane forests expanded and often became continuous; the "seas" that now isolate montane islands were then reduced in extent, and in many cases they were no longer present. The result of this was an effect that was similar to that caused by the lowering of sea level on the Asian continental shelf. As a consequence of the shift downward in elevation of habitats in the terrestrial archipelago, the central Arizona–western New Mexico high country was directly connected with the Rocky Mountains during the Pleistocene (Lomolino et al. 1989) by forest habitat just as Borneo, Sumatra, and Java were connected with southern Asia.

In the Malay Archipelago, Sulawesi (figure 8.2)—called Celebes in the time of Wallace—and adjacent smaller islands are oceanic islands since they are on neither the Asian nor the Australian continental shelf. Even though Sulawesi lies just southeast of Borneo and is geographically much closer to Borneo than Borneo is to the Asian mainland, Sulawesi is separated from Borneo by a deep-water channel (the Makassar Strait); as a result, Borneo and Sulawesi were not connected at any time during the Pleistocene. Although geographically close to Borneo, as a result of its continued isolation Sulawesi is highly distinct biogeographically.

Similarly, in our terrestrial archipelago, the Pinaleño (Graham) Mountains occur just south of the Borneolike central Arizona–western New Mexico high country (as Sulawesi is just southeast of Borneo in the Malay Archipelago). And just as Sulawesi is separated from Borneo by the deep Makassar Strait, the Pinaleño Mountains are separated from the high country to the north by the low-elevation Gila River valley and its desertscrub habitat. As with Sulawesi and Borneo, the Pinaleño Mountains were never connected to the high country during the Pleistocene: the elevation of the floor of the Gila River valley is too low to have ever supported a forest connection (Lomolino et al. 1989).

It would be inappropriate, of course, to push the Malay/Pinaleño analogy beyond reasonable logic. The terrestrial archipelago in Arizona and New Mexico occurs in a very limited geographical region extending though a small range of latitudes; these montane islands, and the distances between them, are relatively small. Unlike the largely tropical Malay Archipelago lying astride the equator, the climate in the terrestrial archipelago of the Pinaleños is entirely temperate or subtemperate. In addition, the Malay Archipelago has a long history of drifting continental plates resulting in changing geographical relationships between islands, and between islands and the mainlands; indeed, geologists have suggested that Sulawesi is a composite island consisting of more than one plate (Audley-Charles 1987).

As a consequence of such differences, lists of taxa at any taxonomic level for each montane island in the terrestrial archipelago are much shorter, and biologists have identified relatively few endemics (i.e., the diversity is lower relative to the Malay Archipelago). In addition, the amount of significant geological time involved in shaping these biotas is probably considerably less. In short, the terrestrial system is orders of magnitude less complex.

However, given the similarities and differences enumerated, several interesting and significant questions arise:

A Subtraction-Transition Zone?

The Malay Archipelago provides the classical example of Darlington's Subtraction-Transition Zone between two major biogeographical regions (Darlington 1957)—Asia to the northwest and the Australian Region to the southeast. Within this zone there is a transition of northern and southern faunas, and the centralmost islands contain progressively fewer taxa (subtraction). Does the terrestrial archipelago also exhibit a subtraction-transition between its two major biogeographical regions (the two "mainlands"), and if so, for which taxa? Or is the transition in this terrestrial archipelago more one of elevational "interdigitation," in which (on mountains with sufficient elevation) species associated with the northern forests are simply superimposed on species associated with the southern woodlands? And of special significance, if progressive subtraction (with increasing distance from the nearest mainland) does occur, how much of this is the result of a negative correlation between distances from the nearest mainland and island areas?

A Terrestrial Wallace's Line?

In the Malay Archipelago, Sulawesi was the island of special interest to Wallace; it was a continual enigma to him. In drawing what came to be called Wallace's Line, representing the most northwestern influence of the Australian Region, he first drew the line northwest of Sulawesi in the Makassar Strait, then later to the southeast—Sulawesi was close to Borneo geographically, yet it was significantly different biologically. His final decision was to draw the line southeast of Sulawesi and in doing so to include this oceanic island with those on the continental shelf of the Asian mainland. Does such a line also occur in the terrestrial archipelago?

Several years ago Charles H. Lowe and Howard K. Gloyd (then of the faculty of the University of Arizona's Zoology Department) suggested that the Interstate-10 highway, which cut east and west through southeastern Arizona south of the Pinaleños, would serve as an appropriate Wallace's Line analogy. A major problem, however, in the designation of any biogeographical-ecological line is the lack of agreement about what such a line should represent. This, I think, has been a major cause of much of the drawing and redrawing of lines in the Malay

Archipelago since the time of the original publication of the results of Wallace's work. In any case, prior to any decision as to the most suitable definition for the situation in the terrestrial archipelago and only on the basis of an intuitive analysis of inadequate data, I tentatively suggest that a line drawn north of the Pinaleños would be more appropriate. I base this suggestion on the strong Madrean influence in these mountains and on what appears to be a major role played by the Gila River basin as a barrier to vertebrate dispersal. But has the Gila River basin been a stronger filter barrier than the Makassar Strait in the Malay Archipelago?

Dispersal Routes?

Although patterns of vicariance undoubtedly occur, is the distribution of certain montane taxa within the terrestrial archipelago also the result of dispersal? Will we be able to recognize routes along which dispersal has occurred that are similar to those demonstrated for plants in the Malay Archipelago (van Balgooy 1987)? Will dispersal patterns show equally significant effects from north to south as in the opposite direction?

Outliers?

In their relationship to the Malay Archipelago, the Philippine Islands are outliers. Their source of species comes almost entirely from the Asian mainland, with little or no influence from the Australian Region (Heaney 1986, 1993; Heaney and Rickart 1990). The string of sky islands in New Mexico east of the Rio Grande may be Philippine analogues in the terrestrial archipelago. Is the source of species on these islands almost entirely from the Rocky Mountain mainland to the north and not from the Sierra Madres to the south (see Chapter 9)?

The proposed analogy and the questions that have been suggested to test it demonstrate that the natural experiments provided by mountain islands continue to be important and stimulating sources of biogeographical investigation. I suspect that of all the mountain islands in southeastern Arizona, the Pinaleños (the potential Sulawesi surrogate) will prove to be one of the most interesting.

Afterthoughts

Scientific considerations aside, people in general—not just ecologists and biogeographers—typically place high value on islands. Perhaps this perception has always been the case throughout human history; certainly it began long before scientific interests arose. Islands provide economic, recreational, and esthetic values. In addition, islands have strong emotional value because they can be special places of mystery and excitement and promise the lure of exploration and adventure. When Wallace first crossed the Makassar Strait from Borneo, with Sulawesi rising before him out of the mist, I doubt his thoughts were tightly focused on natural selection, Wallace's Line, or his future prominence as the "father of biogeography." He was probably not thinking about how to measure species richness or how to take the requisite steps necessary to make biogeography a science in its own right. Much more important to Wallace at the moment was the promise of mystery, adventure, and discovery provided by that intensely verdant island in the distance. Darwin undoubtedly felt similar impulses when he first sighted the Galapagos Islands.

Most of us feel comparable emotions when we see a tall and majestic mountain island, such as the Pinaleños, rising upward from semidesert grassland. If we had acknowledged these feelings in ourselves—while recognizing the awful damage to this emotional value that human activities have wrought on so many islands (including those of the Malay Archipelago)—we would not then have been surprised that controversial decisions about the use and conservation of the Pinaleño Mountains would bring fierce and confrontational reactions on all sides.

References

Audley-Charles, M. G. 1987. Dispersal of Gondwanaland: Relevance to evolution of the Angiosperms. Chapter 2 in *Biogeographical Evolution of the Malay Archipelago,* ed. T. C. Whitmore. Clarendon Press, Oxford, U.K.

Brown, D. E., and C. H. Lowe. 1983. *Biotic Communities of the Southwest.* General Technical Report RM-78. U.S. Department of Agriculture, Forest Service, Rocky Mountain Forest and Range Experiment Station, Ft. Collins, Colo.

Brown, J. H. 1971. Mammals on mountain tops: Non-equilibrium insular biogeography. *American Naturalist* 105:476–78.

Brown, J. H. 1978. The theory of insular biogeography and the distribution of boreal birds and mammals. *Great Basin Naturalist Memoirs* 2:209–27.

Darlington, P. J., Jr. 1957. *Zoogeography: The Geographic Distribution of Animals*. Wiley, New York.

Findley, J. S., and W. Caire. 1977. The status of mammals in the northern region of the Chihuahuan Desert. In *Transactions of the Symposium on the Biological Resources of the Chihuahuan Desert Region,* ed. R. H. Wauer and D. H. Riskind, pp. 127–39. National Park Service Transactions and Proceedings Series Number 3. GPO, Washington, D.C.

Hagmeier, E. E., and C. C. Stults. 1964. A numerical analysis of the distributional patterns of North American mammals. *Systematic Zoology* 13:125–55.

Heaney, L. R. 1986. Biogeography of mammals in Southeast Asia: Estimates of rates of colonization, extinction, and speciation. *Biological Journal of the Linnean Society* 28:127–65.

Heaney, L. R. 1993. Biodiversity patterns and the conservation of mammals in the Philippines. *Asia Life Sciences* 2:261–74.

Heaney, L. R., and E. A. Rickart. 1990. Correlations of clades and clines: Geographic, elevational, and phylogenetic distribution patterns among Philippine mammals. In *Vertebrate Biogeography and Speciation in the Tropics,* ed. G. Peters and R. Hutterer, Museum of Alexander Koenig, Bonn, West Germany.

Lomolino, M. V., J. H. Brown, and R. Davis. 1989. Island biogeography of montane forest mammals in the American Southwest. *Ecology* 70:180–94.

MacArthur, R. H., and E. O. Wilson. 1967. The theory of island biogeography. Princeton University Press, Princeton, N.J.

McLaughlin, S. P. 1989. Natural floristic areas of the western United States. *Journal of Biogeography* 16:239–48.

van Balgooy, M. M. J. 1987. A plant geographical analysis of Sulawesi. Chapter 8 in *Biogeographical Evolution of the Malay Archipelago,* ed. T. C. Whitmore. Clarendon Press, Oxford, U.K.

CHAPTER NINE

Biogeographic Controls of Avifaunal Richness in Isolated Coniferous Forests of the U.S.-Mexican Borderlands

Frederick R. Gehlbach

Because large patches of habitat house more species than small patches, the islandlike coniferous forests at the tops of isolated mountains are of biogeographic interest. Usually these forests are surrounded by woodlands and often by grasslands and deserts in the southwestern United States and northern Mexico, so biologists presume that most animals associated with coniferous-forest structure cannot disperse among the mountain tops. Researchers believe that cooler and wetter climates of the past (Pleistocene ice ages) displaced the forests downward, connecting some mountains and permitting dispersal among nearby ranges, but not distant ones that remained disconnected or were too far from sources of colonists. When forests retreated upslope with the onset of warmer and dryer climates, some resident species may have died out due to resource shortages caused by the reduced habitat area (Chapter 8; Gehlbach 1981).

The Pinaleño (Graham) Mountains provide a large island of forest habitat and thus should house many forest-associated animals. Also, they are relatively close to an even larger upland mass, the Mogollon Plateau and its mountains, so they might be expected to harbor more species on the basis of proximity and greater likelihood of dispersal. Biologists have tried to define which factor—habitat area or isolation distance—is most important to species richness. Some studies (e.g., Wauer and Ligon 1977) suggest that both determine montane avifaunas in the U.S.-Mexican border country, and others (e.g., Lomolino et al. 1989) indicate the same thing about isolated forest mammals, including those of the Pinaleños. However, some research implicates habitat area more than distance in the control of bird species richness on mountain islands north of the Pinaleños (Brown 1971; Johnson 1975; Thompson 1978).

Between 1954 and 1987 I studied breeding birds in mature coniferous forests on mountain islands in southeastern Arizona and adjacent New Mexico, as well as eastward into southern New Mexico and trans-Pecos Texas. The areas studied included the Chisos, Davis, Guadalupe, Organ, and Sacramento mountains east of the Rio Grande lowlands and the Animas, Chiricahua, Huachuca, Peloncillo, Pinaleño, Santa Catalina, and Santa Rita mountains west of the Rio Grande. My primary concern was the nesting biology of selected species, particularly owls, but I also wanted to know why certain species occupied only a few islands despite the regionwide similarity of forest structure. This essay provides some provisional answers, with a special look at the Pinaleño Mountains in view of the current controversy about use of the forest land for animals or telescopes.

Field Methods

Species with breeding niches in coniferous forest habitats (Alexander et al. 1984) of ponderosa pine, mixed conifer, and spruce-fir were studied. I made a census of three or more sites on each mountain island by recording evidence of nesting (nests, fledglings, territoriality) in at least three different years during the months of May through July, for a minimum of three days per site per year (except the Animas and Organ ranges, visited only once for two days each, for which I rely on Niles 1966 and Baltosser [personal communication 1989] for more information). In the Guadalupe, Chiricahua, and Huachuca Mountains I spent continuous one- to two-month periods over a span of two to five years. Moreover, I have observed coniferous-forest birds in both the Arizona and New Mexico portions of the Mogollon Plateau, in the Sierra Madres of Mexico, and elsewhere in North and Central America from Alaska to Guatemala.

Over the 33-year duration of my surveys, major changes have occurred in both landscapes and the distribution of bird populations. For example, the general climate was very dry in the early 1950s but much wetter after 1957, lumbering and especially recreational use of the isolated forests increased, and certain birds appeared or disappeared from certain islands (Gehlbach 1981). Magnificent hummingbirds now breed in the northernmost ranges (Pinaleños, Sacramentos—indeed, also on the Mogollon Plateau) but were absent there at the start, while buff-breasted flycatchers declined or disappeared from northern areas like the Pinaleños. These are just two illustrations of change in the avian populations; but for this study I include all species that I personally saw nesting or defending territories in any year (table 9.1).

Preliminary Appraisals

I considered the closest coniferous forests with at least five times the combined forest area of the islands (1,633 km^2) to be the sources of avian colonists for the islands I sampled. These forests are located on the Mogollon Plateau and its included mountains to the north plus the Sierra Madres, both Oriental and Occidental, to the south. Then I examined the North American breeding ranges of each recorded bird and identified three major distributional patterns (avifaunal components; table 9.1). Boreal-Rockies species inhabit boreal North America and the Rocky Mountains and similar western ranges, plus the northern Sierra Madres in some cases. Madrean birds nest in one or both of the Sierra Madres, the high ranges of the Mesa Central of Mexico, and sometimes southward into Middle America. Rockies-Madrean birds nest in the Rocky Mountains, Sierra Madres, and adjacent mountains.

Since the five eastern islands in New Mexico and Texas appeared to be widespread and farther from the two source areas than the seven western islands, I made some limited comparisons of their attributes, including the three avifaunal components, to determine whether I should conduct separate investigations of the habitat-area versus isolation-distance questions. First I ran a principal components analysis, using the number of birds in each assemblage per mountain island as a variable, and discovered that the eastern islands separated from all but the Animas and Peloncillo islands of the western group on the basis of their relatively depauperate Boreal-Rockies and Madrean assemblages (65% variance explained). The lowlands of the Rio Grande Valley, therefore, appear to divide two different biogeographic regions, at least in part.

For further verification, I compared the average physical and biotic features of the two island groups (data were linearized in all statistical analyses, using appropriate logarithmic or arcsine transformations). This time I considered each avifaunal component in terms of percent saturation (relative species richness), which is the number of nesting species per island divided by the number of all possible nesting species (table 9.2). And I added a third potential constraint on avifaunal richness, the number of mountain islands of the same or larger size between

Table 9.1 Biogeographic Assemblages of Coniferous Forest Birds That Nest in Colonial Source Areas

Boreal-Rockies	*N*	Sierra Madrean	*N*
Cordilleran flycatcher (*Empidonax occidentalis*)	12	Magnificent hummingbird (*Eugenes fulgens*)*	10
Western tanager (*Piranga ludoviciana*)*	11	Greater pewee (*Contopus pertinax*)*	7
Hermit thrush (*Catharus guttatus*)*	10	Red-faced warbler (*Cardellina rubrifrons*)*	7
Sharp-shinned hawk (*Accipiter striatus*)*	9	Olive warbler (*Peucedramus taeniatus*)*	6
Red crossbill (*Loxia curvirostra*)*	8	Yellow-eyed junco (*Junco phaeonotus*)*	6
Yellow-rumped warbler (*Dendroica coronata*)*	7	Buff-breasted flycatcher (*Empidonax fulvifrons*)	5
Pine siskin (*Carduelis pinus*)*	7	Mexican chickadee (*Parus sclateri*)	3
Red-breasted nuthatch (*Sitta canadensis*)*	6	Colima warbler (*Vermivora crissalis*)	1
Mountain chickadee (*Parus gambeli*)*	6	Thick-billed parrot (*Rhynchopsitta pachyrhyncha*)	0
Goshawk (*Accipiter gentilis*)*	6	White-eared hummingbird (*Hylocharis leucotis*)	0
Northern saw-whet owl (*Aegolius acadicus*)*	5	Eared trogon (*Euptilotis neoxenus*)	0
Ruby-crowned kinglet (*Regulus calendula*)*	4	Black-headed siskin (*Carduelis notata*)	0
Orange-crowned warbler (*Vermivora celata*)*	4	Striped sparrow (*Oriturus superciliosus*)	0

Golden-crowned kinglet (*Regulus satrapa*)*	4
Red-naped sapsucker (*Sphyrapicus nuchalis*)	2
Dark-eyed junco (*Junco hyemalis*)	2
Olive-sided flycatcher (*Contopus borealis*)	2
MacGillivray's warbler (*Oporornis tolmiei*)	1
Blue grouse (*Dendragapus obscurus*)	0
Williamson's sapsucker (*Sphyrapicus thyroideus*)	0
Three-toed woodpecker (*Picoides tridactylus*)	0
Dusky flycatcher (*Empidonax oberholseri*)	0
Clark's nutcracker (*Nucifraga columbiana*)	0
Evening grosbeak (*Coccothraustes vespertina*)	0
Pine grosbeak (*Pinicola enucleator*)	0

Rockies-Madrean	*N*
Broad-tailed hummingbird (*Selasphorus platycercus*)	12
Steller's jay (*Cyanocitta stelleri*)*	11
Pygmy nuthatch (*Sitta pygmaea*)*	11
Flammulated owl (*Otus flammeolus*)*	10
Grace's warbler (*Dendroica graciae*)*	10
Spotted owl (*Strix occidentalis*)*	8
Brown creeper (*Certhia americana*)*	8
Northern pygmy owl (*Glaucidium gnoma*)*	7
Townsend's solitaire (*Myadestes townsendi*)	1

Note: Species nest on the Mogollon Plateau, Arizona–New Mexico, and/or northern Sierra Madres Oriental and Occidental, Mexico. These species also nest or, because of suitable habitat, could nest on 12 mountain islands between these source areas. *N* indicates number of mountain islands occupied, * indicates nesting in the Pinaleño Mountains.

Table 9.2 Physical and Biotic Features of Coniferous Forest and Its Breeding Avifauna on Mountain Islands East and West of the Rio Grande Lowlands

Features	East ($N = 5$)	West ($N = 7$)
Coniferous forest area (km²)	78	57
Distance to Mogollon Rim, Arizona–New Mexico (km)	1,007	511*
Distance to nearest Sierra Madre, Mexico (km)	6,382	65*
Mountain island "steps" to Mogollon Rim (*N*)	4	2*
Mountain island "steps" to Sierra Madres (*N*)	4	3
Avifaunal saturation (%)		
All possible species	31	49
Boreal–Rocky Mountains species	33	40
Rocky Mountains–Sierra Madrean species	44	57
Sierra Madrean species	9	64*

Note: Mountain islands are in Arizona, New Mexico, and Texas. Saturation percentages are number of actual nesting species divided by number of all possible nesting species. *N* indicates number of mountain islands in each group, * indicates significant difference ($p < 0.05$).

each sampled island and each "mainland" source area, because more island steps (or stopover habitats) might offset an isolation-by-distance influence on colonization.

The results suggest that the eastern islands have more high-elevation and, hence, forest area than the western islands; but variation in the sample, which includes both the largest (Sacramentos, 1,350 km²) and the smallest (Organs, 8 km²) ranges, makes the statistical comparison insignificant. Nevertheless, the eastern islands certainly are farther from both northern and southern sources of colonists, although they have more island steps between them and the Mogollon Plateau than do the western islands. Yet the eastern island group tends to have fewer species than the total possible in each avifaunal assemblage, and the difference in Madrean birds is significant. Overall, the eastern group houses 18% fewer species, apparently because it is farther from mainland sources (table 9.2).

In the analyses that follow I make the reasonable assumption, based on the mountain-mass phenomenon (Gehlbach 1981), that a larger island means more coniferous forest habitat. Also, I have found that isolation distance correlates with the number of island steps to both the Mogollon Plateau ($r = 0.99$) and the Sierra Madre Oriental ($r = 0.99$) in the eastern island group, though not among the western islands ($r < 0.56$, ns), so I have chosen to simplify matters by omitting further analyses involving island steps. Island (habitat) area does not correlate with isolation distance or island steps in either group ($r < 0.68$, ns), but I used simple relationship analyses (linear regressions) rather than a more complex procedure (multiple regression) to focus individually on possibly different biogeographic patterns of habitat as opposed to distance.

Major Patterns

Therefore, for each group of mountain islands, I regressed percent saturation of the total coniferous-forest avifauna and its two most sizeable components (Boreal-Rockies and Madrean; table 9.1) on habitat area and, separately, on isolation distance. In view of the preliminary appraisals, I expected that different patterns might appear for the eastern versus western islands, that isolation could be more important in the east, and that the Madrean birds in particular ought to illustrate major biogeographic differences. I also considered the possibility of an isolation threshold, some distance beyond which the birds have not dispersed because the necessary coniferous forest did not exist in their history.

Because my sample sizes were small, I was less interested in the statistical power of each landscape factor to predict avifaunal richness than in the possibility of identifying different eastern and western patterns. Thus, I was surprised to see similarly strong positive relationships between total species saturation and habitat area in both island groups but not as amazed that saturation seems unrelated to isolation distance (figure 9.1). Perhaps all avifaunal components are similarly influenced by area but not by distance, as indicated by the different slopes for the eastern and western islands (figure 9.1). This possibility required further investigations at the level of individual assemblages.

In the first analysis, Boreal-Rockies birds respond positively to habitat area both in the east ($r^2 = 0.76$, $p = 0.05$) and in the west ($r^2 = 0.74$, $p = 0.01$), but isolation distance operates somewhat differently. On the eastern islands, distance from neither the Mogollon Plateau nor the Sierra Madre Oriental is influential ($r^2 < 0.26$, ns). Among the western islands, by contrast, distance from the Mogollon Plateau is unimportant ($r^2 = 0.15$, ns), but Boreal-Rockies birds respond positively to distance from the Sierra Madre Occidental ($r^2 = 0.73$, $p = 0.01$). Apparently, the northern source of supply is sufficiently close to all western islands, although islands farthest from the southern influence are somewhat better represented by Boreal-Rockies birds. Perhaps because these northern islands have fewer Madrean species, the Boreal-Rockies birds have less competition and can dominate the avifauna.

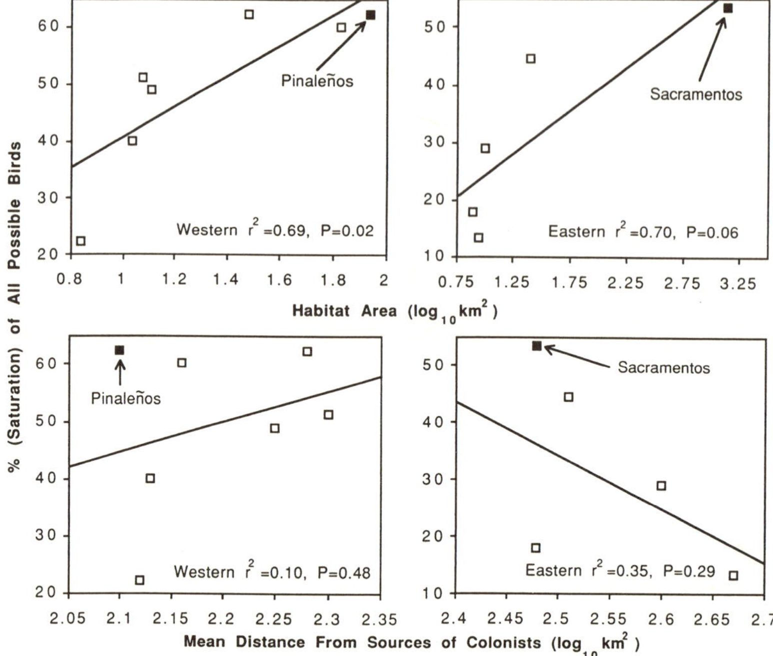

Figure 9.1 Percent saturation (known breeding species/all possible breeding species) of isolated coniferous-forest avifaunas in relation to habitat area and mean distance from sources of colonists on the Mogollon Plateau of Arizona–New Mexico and the Sierra Madres Oriental and Occidental of Mexico. Seven avifaunas west of the Rio Grande lowlands are compared to five others east of this biogeographical dividing line. The data are arcsine (percentages) or $\log_{10}$-transformed, and graph axes are derived accordingly.

Turning to the Madrean element, I found that habitat area has no influence on it in the eastern islands ($r^2 = 0.17$, ns); instead the Madrean group is negatively constrained by distance from the Sierra Madre Oriental ($r^2 = 0.69$, $p = 0.07$). On the west neither area nor distance to the Sierra Madre Occidental determines avifaunal saturation ($r^2 < 0.37$, ns); although if Madrean species are better represented, so are Boreal-Rockies species ($r = 0.72$, $p = 0.06$). This is not the case on the eastern islands ($r = -0.67$, ns), further indicating that distance has been a major factor for these islands. Perhaps 800 km is near the threshold for complete isolation, since this is the approximate distance between the Organ Mountains, the only range totally lacking Madrean species, and the northernmost Sierra Madre Oriental.

I next explored the question of which assemblage dominates the isolated avifaunas. By regressing total avifaunal size on the number of species in each, I noted that the Boreal-Rockies group is a very strong and positive determinant both east and west ($r^2 = 0.96$, 0.88, respectively; $p < 0.001$). The Madrean group follows suit on the western islands ($r^2 = 0.70$, $p = 0.01$) but shows an insignificant negative pattern in the east ($r^2 = 0.06$, ns). To be certain that I did not overlook a possible Rockies-Madrean contribution, I analyzed it in the same fashion and found it well represented both east and west ($r^2 = 0.97$, 0.82, respectively; $p < 0.02$).

The preceding analyses indicated that coniferous-forest avifaunas both east and west of the Rio Grande lowlands are dominated by species of northern biogeographic affinities. Furthermore, Boreal-Rockies birds respond strongly and positively to the amount of habitat, whereas Madrean birds do not appear to be controlled by habitat area, possibly because they are more dry-adapted and tolerant of bordering pine-oak-juniper woodlands. On the eastern islands, quite distant from their homeland, Madrean birds either have not colonized or have disappeared from avifaunas coincident with the degree of forest isolation. Thus, distance does appear to be a threshold factor, but this factor holds chiefly for birds of southern biogeographic affinities east of the Rio Grande lowlands.

Species Relationships

In exploring the essential question of this study further, I also attempted to discover how widespread species respond to habitat area versus isolation distance. By selecting nine birds from the Boreal-Rockies and Madrean groups that nest on some but not all mountain islands, both east and west of the Rio Grande, I studied influences of the two landscape factors relatively uncomplicated by the east-west biogeographic differences. Additionally, I took into account that species might disappear from or never colonize an island because of competitors as suggested by the offset distributions of, for example, dark-eyed and yellow-eyed juncos or mountain and Mexican chickadees (Hubbard 1978; Monson and Phillips 1981). I assayed the influence of feeding-guild membership, therefore, by logistically regressing the presence or absence of each bird on guild size as well as on area and distance.

In this analysis, habitat area emerges as the most pervasive factor, but only for the Boreal-Rockies contingent. Habitat area explains

35–62% of occurrence and is a significant or near-significant ($p = 0.07$) positive influence on all seven Boreal-Rockies species (northern saw-whet owl, ruby-crowned kinglet, red-breasted nuthatch, mountain chickadee, yellow-rumped and orange-crowned warblers, and pine siskin). Guild size is also a positive determinant in four cases ($r^2 > 0.42$, $p < 0.02$), while distance to either source of colonists is not implicated at all. Conversely, the magnificent hummingbird and red-faced warbler of the Madrean group are not influenced by area but instead are negatively controlled by mean distance from both Sierra Madres ($r^2 > 0.52$, $p < 0.008$). For these species, guild size is a factor (positive) only for the warbler ($r^2 = 0.42$, $p = 0.02$).

Although large patches of habitat are most likely to intercept and retain species and house all guild members with minimum adverse consequences of competition for space and other resources, this factor affects only Boreal-Rockies birds. Birds of this assemblage could be larger and hence more mobile than Madrean birds, more likely to reach islands regardless of distance; but, as a disadvantage, they would require more space. When I analyzed this possibility I determined no difference in mean weight ($t = 0.2$, ns) and no general correlation between body weight and number of islands occupied among Boreal-Rockies ($r = 0.02$) or Madrean species ($r = 0.08$). Nevertheless, body size is positively correlated with number of islands occupied among the 16 insect-eating species of both Boreal-Rockies and Madrean groups ($r = 0.61$, $p = 0.01$), which are smaller (<31 g) than the remaining 10 vertebrate and seed-eating species ($t = 1.9$, $p = 0.05$).

Overall, this analysis indicated that widely distributed species have reached or persisted in larger habitats, but the smallest (insectivorous) birds and Boreal-Rockies species in particular are constrained by habitat area. Small Madrean birds, by contrast, are more seriously affected by isolation, like all birds of that assemblage. The two ultimate landscape factors, habitat area and isolation distance, seem to be more potent than guild size, a proximate biological factor, in determining how widespread a species is. Nevertheless, if an island's forest habitat contains one member of a guild, it seems likely to house others in many (56%) cases.

Since my fieldwork focused on the owl guild, which includes the spotted owl—recently considered a threatened species in the Southwest (Ganey and Balda 1989)—a few specific observations may be important to conservation in terms of habitat area. This species nests on each sampled island except the Chisos, Davis, Organs, and Peloncillos (which have the four smallest forest patches, all less than 10 km²). Yet on some islands like the Guadalupes, the spotted owl is chiefly a cliff-crevice nester in deep canyons amidst riparian vegetation, not a frequent user of old stick nests or tree cavities in continuous coniferous forest. Habitat area is just a possible minor influence on its presence or absence ($r^2 = 0.28$, $p = 0.07$); isolation distance is unimportant ($r^2 < 0.22$, ns).

Instead, the four-member owl guild as a whole responds positively to habitat area ($r^2 = 0.50$, $p = 0.009$), not to distance from either source of colonists ($r^2 < 0.17$, ns). The spotted owl eats small and medium-size rodents, the saw-whet eats small rodents and a few birds, the northern pygmy owl also takes small rodents plus many small birds and insects, and the flammulated owl consumes insects almost entirely. Some rodents girdle seedling and sapling conifers and eat their leaves and seeds, some birds eat conifer seeds, and certain insects also injure or kill conifers. The owls may help to control conifer mortality by regulating a diverse array of conifer predators and, hence, contribute to larger and possibly more stable forest patches. Perhaps the owl guild promotes its coexistence with a variety of other organisms (Mannan 1980).

Pinaleño Forest Birds

Next to the Sacramento Mountains, the Pinaleños contain the largest area of coniferous forest, but they are western and, hence, are closer to both northern and southern sources of colonists (figure 9.1; table 9.2). Thus, it is not surprising that the Pinaleños together with the Santa Catalinas have the richest coniferous-forest avifauna of all the western islands. This avifauna comprises 28 (80%) of the 35 species that nest on the sample areas of this study and 60% of the 47 actual and potential colonists (table 9.1). By comparison with the six other western islands, the more northerly Pinaleños house 23 percent more Boreal-Rockies species together with similar fractions of the other species assemblages and are quite northern in biogeographic aspect (figure 9.2).

Only the Sacramentos are more saturated with Boreal-Rockies and Rockies-Madrean species, undoubtedly because they are much more massive (15.5 times larger) rather than because they are farther north (by only 0.8 degrees of latitude). But the Pinaleños house six times more Madrean species because of their proximity to the Sierra Madre Occidental, so they have a more even representation of all three biogeographic elements (Chapter 8; figure 9.2). Among particular species only

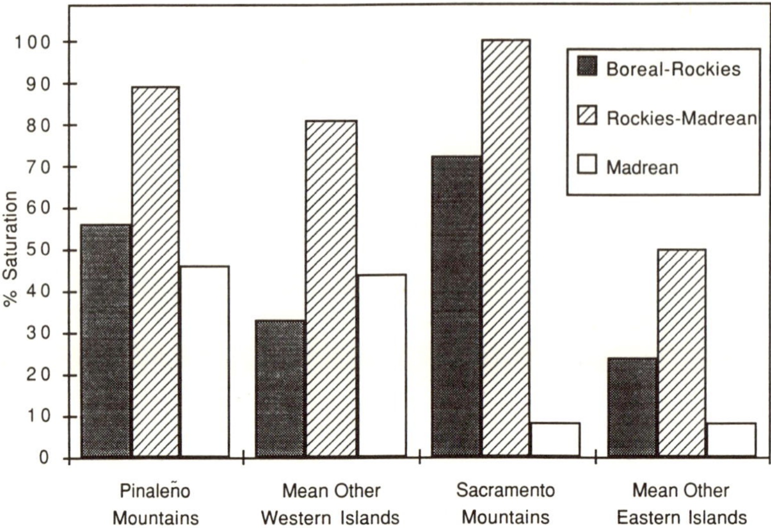

Figure 9.2 Percent saturation (known breeding species/all possible breeding species) of three biogeographically distinct avifaunal elements in the Pinaleño Mountains (87 km²) by comparison with the Sacramento Mountains (1,350 km²) and the means of other mountain islands west and east of the Rio Grande lowlands.

the spotted owl and buff-breasted flycatcher are perhaps unusual; the owl as explained above, and the flycatcher because it inhabits only four other mountain islands. Of similarly restricted birds, only the Colima warbler, Mexican chickadee, and Townsend's solitaire are more confined (table 9.1).

Surely the avifauna of the Pinaleños is especially rich because of the large habitat area of this mountain island, not its small degree of isolation (figure 9.1). Birds of the widespread Boreal-Rockies and Rockies-Madrean assemblages, missing from the Pinaleños, that nest on one or more other mountain islands are the red-naped sapsucker, olive-sided flycatcher, Townsend's solitaire, MacGillivray's warbler, and dark-eyed junco. But these species do not inhabit any of the western islands I sampled, suggesting that some distributional accident has limited them to the eastern group. It seems unlikely that they would have colonized and then disappeared from all the western islands. In sum, coniferous forest birds of the Pinaleños are somewhat more boreal than, but otherwise representative of, forested island avifaunas along the U.S.-Mexican border west of the Rio Grande.

Practical Biogeography

Any scheme of natural resource management begins with goals stressing features that are known to be important and includes priorities that take into account conflicting interests. The preservation of biodiversity is one unquestioned goal of conservation biologists, and, if resource managers must choose one species over another, threatened or endangered species have top priority. My findings suggest that one way to maintain biodiversity (avifaunal richness) on isolated mountaintops is to protect habitat area and that unusual or threatened species like the spotted owl may be members of guilds that respond collectively to increased habitat. Furthermore, the preservation of biodiversity might be self-promoting through the ecological regulatory effects of guilds once habitat is preserved.

Distance from sources of colonists is the other large-scale landscape factor that determines biodiversity on mountain islands, but it is an academic consideration in that we cannot reduce distances. We can, however, elect to save all forested mountaintops so that isolation might be offset by intermediate steps or stopovers for dispersing species. If we must choose between protecting eastern or western islands, the eastern ones deserve first consideration, since their Madrean avifaunas are especially limited by isolation. From the biogeographic perspective of relatively isolated eastern avifaunas, preserve status for the Davis Mountains makes good sense, since that range is the only forested step between the Chisos and Guadalupe mountains, which are already protected in national parks.

Some environmentalists have questioned whether the Pinaleños encompass sufficient habitat area for the threatened spotted owl and its endangered prey, the Mt. Graham red squirrel. Yet this range houses one of the two richest avifaunas among the western islands in which the extent of habitat appears to be the main reason for avifaunal richness. The simple answer to preserving biodiversity in the Pinaleños would be to disallow observatory construction, which removes acreage from coniferous forest that is the preferred habitat of the spotted owl and Mt. Graham red squirrel. But the answer is not simple, since the Santa Catalina Mountains have exactly the same species of nesting birds in only about one-third the area (30 km^2 versus 87 km^2). Does this comparison suggest that some part of the Pinaleño forest can be used for an observatory without negatively affecting these protected species?

I approached this question by determining the asymptote of total

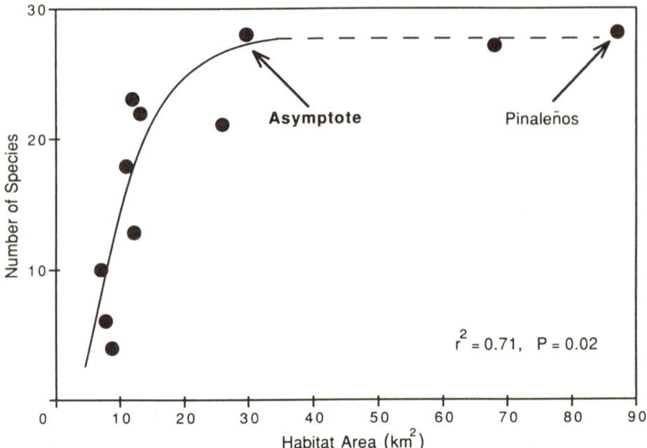

Figure 9.3 Breeding species richness of the coniferous-forest avifauna relative to habitat area on 11 mountain islands (Sacramento Mountains with 28 species and 1,350 km² area is omitted). The asymptote or maximum of 28 species corresponds to 30 km² in a 2° polynomial regression.

avifaunal richness relative to habitat area. I was also interested in the owl guild, because it could be a functional indicator of relatively stable habitat. Since the size of this guild is strongly related to total avifaunal richness ($r = 0.96$, $p < 0.001$), I decided to examine the total, because it is prudent to enumerate as many interacting species as possible when so many ecological roles are unknown. An avifauna of 28 species in a habitat area of 30 km² is the asymptote (figure 9.3). Thus, 57 km² of the Pinaleño range does not seem to matter, as none of the 12 mountain islands has more than 28 coniferous-forest species.

However, the real ecological world is not based solely on species richness relative to habitat area. For instance, I have made no allowances for landscape diversity, and some species may require special microhabitats for nesting—like the spotted owl's cliff crevices in deep canyons. Moreover, in implying that 30 km² is adequate for preserving mountaintop avifaunas, I do not allow for natural disturbances such as storms, fire, and bark-beetle outbreaks, the negative effects of which may be reduced or localized in a larger forest. Also, larger habitats mean larger populations per species, each with greater genetic diversity that may preadapt species to environmental change. Finally, any nature preserve is islandlike because of human cultural influences around its edges; hence, outside buffer zones must be designated to reduce deleterious edge effects.

The special lesson of this study is that the fewest coniferous-forest birds will be lost on mountain islands with forests larger than 30 km^2, provided that resource managers mitigate negative factors and establish safeguards such as buffer zones. The largest patches are the best places for cultural installations, if choices can be made among forest areas of different sizes. But the amount of sacrificed forest must be small to minimize adverse consequences for population size. Any new construction area should not exceed the average annual loss in habitat from natural and other cultural factors, some of which can be eliminated (e.g., lumbering) or reduced in extent (e.g., recreation) to compensate for the new loss of habitat. The general lesson from biogeography is that tradeoffs and harmonious mixes are part of every ecosystem (e.g., MacArthur and Wilson 1967; Wilson 1992).

Acknowledgments

I have had many field companions over the years, but I am especially indebted to Nancy Gehlbach. Robert Baldridge, William Baltosser, John Hubbard, and Joe Marshall Jr. helped with the data. Russell Davis, Nancy Gehlbach, and Joe Marshall read the text at one or more stages of its development. Also, Joe and I conversed and corresponded many times about our various experiences with southwestern and Mexican birds and the Pinaleño avifauna. I thank all these folks very much and also appreciate Conrad Istock's willingness to deal forthrightly with the rapprochement of nature and culture.

References

Alexander, B. G., F. Ronco, E. L. Fitzhugh, and J. A. Ludwig. 1984. *A Classification of Forest Habitat Types of the Lincoln National Forest, New Mexico.* General Technical Report RM-104. U.S. Department of Agriculture, Forest Service, Ft. Collins, Colo.

Brown, J. H. 1978. The theory of insular biogeography and the distribution of boreal birds and mammals. *Great Basin Naturalist Memoirs* 2:209–27.

Ganey, J. L., and R. P. Balda. 1989. Distribution and habitat use of Mexican spotted owls in Arizona. *Condor* 91:355–61.

Gehlbach, F. R. 1981. *Mountain Islands and Desert Seas: A Natural History of the U.S.-Mexican Borderlands.* Texas A & M University Press, College Station. (Reprinted 1993.)

Hubbard, J. P. 1978. *Revised Check-List of the Birds of New Mexico.* New Mexico Ornithological Society Publication Number 6. Albuquerque.

Johnson, N. K. 1975. Controls of number of bird species on montane islands in the Great Basin. *Evolution* 29:545–67.

Lomolino, M. V., J. H. Brown, and R. Davis. 1989. Island biogeography of montane forest mammals in the American Southwest. *Ecology* 70:180–94.

MacArthur, R. H., and E. O. Wilson. 1967. *The Theory of Island Biogeography.* Princeton University Press, Princeton, N.J.

Mannan, R. W. 1980. Assemblages of bird species in western coniferous old-growth forests. In *Management of Western Forests and Grasslands for Nongame Birds,* ed. R. M. DeGraff and N. G. Tilghman, pp. 357–68. General Technical Report INT-86. U.S. Department of Agriculture, U.S. Forest Service, Ogden, Utah.

Monson, G., and A. R. Phillips. 1981. *Annotated Checklist of the Birds of Arizona.* University of Arizona Press, Tucson.

Niles, D. M. 1966. *Observations on the Summer Birds of the Animas Mountains, New Mexico.* New Mexico Ornithological Society Publication Number 2. Albuquerque.

Thompson, L. S. 1978. Species abundance and habitat relations of an insular montane avifauna. *Condor* 80:1–14.

Wauer, R. H., and J. D. Ligon. 1977. Distributional relations of breeding avifauna of four southwestern mountain ranges. In *Symposium on the Biological Resources of the Chihuahuan Desert Region,* ed. R. H. Wauer and D. H. Riskind, pp. 567–78. National Park Service, Transactions and Proceedings Series Number 3. GPO, Washington, D.C.

Wilson, E. O. 1992. *The Diversity of Life.* W. W. Norton & Company, New York.

CHAPTER TEN

Local Extinctions and the Biogeographic Dynamics of Boreal Mammals in the Southwest

Bruce D. Patterson

The distinctive montane faunas and floras inhabiting "sky islands" of the American Southwest have long fascinated field biologists. The origin and derivation of these biotas, either by evolution in situ or by colonization from beyond inhospitable deserts, have proven particularly puzzling. Biologists have advanced different and contrasting hypotheses to explain the origin and isolated nature of these distributions. In fact, many hypotheses are compatible with the distribution patterns, but these explanations differ dramatically with respect to how rapidly distributions have shifted. To distinguish among hypotheses, we need to know when plants and animals arrived on mountain ranges and when various isolated populations became extinct. Understanding the distinctiveness and forecasting the fates of isolated montane populations (like those of the Pinaleños) require information on distributional dynamics.

Background on Distributional Dynamics

Joseph Grinnell and Harry Swarth (1913:385) first articulated what has become a widely accepted view of montane biogeography: "The facts show that fewer types have been preserved in one place than another, and further that the following statement of a possible law appears to be justified: *The smaller the disconnected area of a given zone (or distributional area of any other rank), the fewer the types which are persistent therein*" (emphasis in the original). They argued that disjunct bird and mammal faunas of the San Jacinto region of southern California had been derived by means of colonizations during the Pleistocene (the most recent series of global cooling episodes), followed by subsequent local extinctions of intervening lowland populations. Extinctions supposedly occurred as once-continuous Pleistocene distributions became fragmented and isolated by changing Holocene climates.

Since their seminal paper, numerous biologists have invoked little-modified versions of the Grinnell-Swarth hypothesis to explain the currently disjunct distributions of a variety of plants, mollusks, fishes, and mammals in the Great Basin and Southern Rocky Mountains provinces (e.g., Findley 1969; Brown 1971; Metcalf 1977; Smith 1978; Wells 1983; Patterson 1984; Sullivan 1985). Because colonization and extinction operate in sequence, this hypothesis contrasts with equilibrium island biogeography. These "non-equilibrium" cases have become paradigms for how other, less studied, systems became populated, mainly because the Quaternary history of the Southwest is well documented. For example, biologists have widely applied the scenario of widespread colonization during the Pleistocene—and subsequent fragmentation and local extinction during the Holocene (approximately the last 10,000 years)—especially to landbridge islands; it also serves as a model for conservationists studying ecological reserves (e.g., Diamond 1984).

However, an opposing view holds that dispersal among mountain ranges continues to play a major role in shaping the distributions of modern faunas. One study (Davis et al. 1988) showed that significant variation in the species richness of montane faunas in the southern Rocky Mountains can be attributed to their isolation from colonization sources. Another (Lomolino et al. 1989) determined that incidence patterns (i.e., the presence or absence of species on different mountain ranges) were also affected by isolation. Researchers have interpreted these statistical "isolation effects" as evidence for ongoing dispersal through lower-elevation woodland and grassland habitats. A mechanism for producing isolation effects in the fauna as a whole was suggested by detailed natural history studies of selected species (Davis and Dunford 1987; Davis and Brown 1989), which documented historic range expansions by cotton rats and tassel-eared squirrels, respectively. Distributions of montane mammals in the Southwest appeared so dynamic to Mark Lomolino and his coworkers (1989) as to approximate an ongoing equilibrium of colonization and extinction.

Both of these interpretations of Southwestern biological history share many important components and predictions, and few tests or observations can lead to unambiguous rejection of one or the other.

Simple observation is impractical, because dispersal is notoriously difficult to observe and even very low levels of dispersal can have major evolutionary consequences (Wright 1978). That these contradictory alternatives are both viable and descriptive underscores the complexity of historical biogeography and the probable inadequacy of simple, single-factor explanations. However, it also reflects a methodological weakness shared by both alternatives as they have been applied to boreal mammals of the southwestern United States. Researchers have based each viewpoint mainly on analyses of modern distributions, with little or no explicit reference to the historical record. Studies of modern distributions lack a temporal dimension that is essential to understanding the dynamics of distributions.

James Brown's (1971) classic and insightful analysis of mammalian distributions on Great Basin mountain ranges illustrates this limitation. Brown's regression analyses showed that current species richness is strongly correlated with area and not with measures of isolation. However, because mammals restricted to mesic forest habitats do not occur on any of these ranges today, Brown concluded that they had been unable to colonize these ranges during the Pleistocene. Subsequent fossil discoveries in Great Basin lowlands strongly substantiated the importance of postglacial extinction that Brown had inferred from regression analyses (Grayson 1987). However, the fossils did more than this. Extralimital fossils (remains of animals no longer present locally) of mesic-adapted forms provided cogent evidence that the study of modern distributions significantly underestimated the extent of glacial colonizations, and therefore the extent of postglacial extinctions. Several forms strongly associated with cool mesic habitats (e.g., pikas and heather voles: scientific names are provided in table 10.1) apparently became broadly distributed during the Wisconsin glaciation but disappeared from all the isolated ranges during postglacial times.

In this chapter I attempt to add a historical dimension to ongoing biogeographic discussions in the Southwest, using the region's remarkable Quaternary record. At least 360 vertebrate taxa are known from Quaternary deposits at more than 350 sites in the western United States (Harris 1985, 1990). Like most fossil records, fossil data from the Southwest are fragmentary, have large spatial and temporal gaps, and are often only crudely dated (e.g., "late Wisconsin," the last glaciation, or "early Holocene," the postglacial period; see Harris 1985). However, existing records represent most of the montane mammals and many parts of the Southwestern region. Analyses of fossils indicate that postglacial extinctions have played a major role in the historical derivation of boreal biotas in the Southwest. In one thoroughly documented case, the Guadalupe Mountains, colonization during the Recent epoch has had no effect on the complete disappearance of its boreal mammal fauna. In the following discussion I attempt to relate these findings to patterns in modern distributions.

Materials and Methods

Mountain ranges in the southwestern United States are home to a diverse assemblage of mammals derived from various geographic sources and restricted in varying degrees to montane habitats. In the Southwest, montane mammals—those species living in habitats found principally or exclusively on mountain ranges—include species from the Cordilleran, Boreo-Cordilleran, Campestrian, Great Basin, Yuman, and Chihuahuan faunas, as well as autochthons (species that may have originated in this geographic province) and eurychores (too widespread to be associated with any given fauna; Armstrong 1972; Hoffmeister 1986). By definition, each of these faunas presumably had a unique historical derivation, often under distinct environmental conditions. If biologists are to resolve and clarify history through biogeographic analyses, they must *distinguish* the distributional patterns shown by these different faunas (Patterson 1984). Several contributions to the debate over the biogeography of the southwestern United States (Patterson 1980; Davis and Dunford 1987; Davis et al. 1988; Lomolino et al. 1989) have failed to distinguish "boreal" forms (which moved from the north into the study region) from "austral" taxa (which dispersed into the region from the Sierra Madre Occidental of Mexico). As detailed below, most evidence for ongoing colonization and dynamic distributions of Southwestern mammals comes from austral species.

In this essay, reference to boreal montane mammals includes only members of the Cordilleran and Boreo-Cordilleran faunas (Armstrong 1972), which in the Southwest are mainly restricted to habitats found on mountain ranges. All 24 species (listed in table 10.1) have centers of geographic distribution or sister-group relations to the north. This definition excludes several species that naturally occur with these species, as being either not restricted to mountains (deer mouse; *Peromyscus maniculatus*), indigenous with unknown relationships outside the area (western chipmunks *Eutamias quadrivittatus, E. canipes,* and *E. cinereicollis*), or species of Chihuahuan origin (Abert's squirrel; *Sciurus aberti*).

I based the Quaternary records of boreal species in the Southwest on Arthur Harris's (1985) compilation. Given depositional and taphonomic (fossil preservation) differences among species and sites, I could not estimate distributional range during the Pleistocene from the area between fossil-bearing sites. Instead, I used representation among sites referred to the Wisconsin glacial episode of the Pleistocene (roughly 80,000 to 10,000 years ago; Harris 1985), applying three related indices: the total number of sites in the western United States at which investigators have recovered fossils of each species, the number of these sites in the Southwest (Colorado, Utah, Arizona, New Mexico, and Trans-Pecos Texas), and the number of extralimital sites (those lying beyond the current distributional limits of each species). I also derived a corresponding measure of modern distribution, the number of mountain ranges presently supporting that species from previous studies (Armstrong 1972; Patterson 1984). Body size for each taxon (weights, in g) are averages taken from the literature or from specimens in the Field Museum's mammal collection. I took habitat data for species directly from the work of Stephen R. Bernard and Kenneth F. Brown (1977); these researchers assessed distributions of western vertebrates among A. W. Küchler's (1975) associations of potential natural vegetation. I treated each vegetational association as a separate variable with two states, present or absent, arbitrarily interpreting their term *wide range* to include all associations. Habitat relationships among mammals were assessed using Jaccard's similarity coefficient, which does not tally negative matches and is thus appropriate for use with highly categorized data (Sneath and Sokal 1973).

The Distribution of Pleistocene Fossils

Most (92%) of the boreal taxa inhabiting mountain ranges in the southwestern United States are represented by late Pleistocene fossils at one or more western localities (table 10.1). Of these, only 5 species (23%) have not been identified in Southwestern deposits. Extralimital records exist for more than half (54%) of the 24 boreal mammal species, documenting the greater extent of their geographic ranges during the late Pleistocene. All but one (92%) of the taxa with extralimital records are known from fossil deposits near the southern periphery of the study area. In concert with modern distributions, Quaternary records document dramatic latitudinal expansions of ranges during glacial episodes, resembling those previously documented for plants (e.g., Van Devender and Spaulding 1979; Spaulding et al. 1983). Compare in

Table 10.1 Distributions of Boreal Mammal Species Now and During the Quaternary

Species	Current Ranges	Quaternary Fossil Distribution			
		Total Fossils	Total Southwestern Fossils	Positively Identified Fossils	Extralimital Fossils
Sorex monticolus, montane shrew	26	8[b]	8	6	8
Microtus longicaudus, long-tailed vole	22	11	6	4	6
Tamiasciurus hudsonicus, red squirrel	22	3	3	3	3
Eutamias minimus, least chipmunk	12	0[c]	0	0	0
Spermophilus lateralis, golden-mantled ground squirrel	12	11	4	4	1
Thomomys talpoides complex, northern pocket gophers	12	18	9	9	8
Neotoma cinerea, bushy-tailed woodrats	11	33[d]	25	22	22
Sorex palustris, American water shrew	10	4[b]	2	2	1
Clethrionomys gapperi, Gapper's red-backed vole	10	4	0	0	0
Marmota flaviventris, yellow-bellied marmot	8	36	25	24	21
Sorex nanus, dwarf shrew	8	2[b]	2	1	2
Zapus princeps, western jumping mouse	8	0	0	0	0

Microtus montanus, montane vole	7	15	5	3	4
Mustela erminea, ermine	6	5	1	1	0
Ochotona princeps, Rocky Mountains pika	6	20	3	3	0
Sorex cinereus, masked shrew	5	7[b]	6	6	5
Lepus americanus, snowshoe hare	5	3	0	0	0
Martes americana, pine marten	4	3	2	2	0
Phenacomys intermedius, heather vole	4	9	3	3	1
Gulo gulo, wolverine	3	6	2	2	3
Lynx canadensis, Canadian lynx	3	3	0	0	0
Sorex hoyi, pygmy shrew	1	1[b]	0	0	0
Eutamias umbrinus, Uinta chipmunk	1	1[c]	0	0	0
Spermophilus elegans, elegant ground squirrel	1	13	5	4	3

Note: Current distribution among mountain ranges as in Patterson 1984.

[a] Confirmed (not referred or questionable) identifications within the Southwest region.

[b] Does not include 5 records, 2 in Southwest, of *Sorex* sp.

[c] Does not include 15 records, 8 in Southwest, of *Eutamias* sp.

[d] Does not include record identified as either *Neotoma cinerea* or *Neotoma mexicana*.

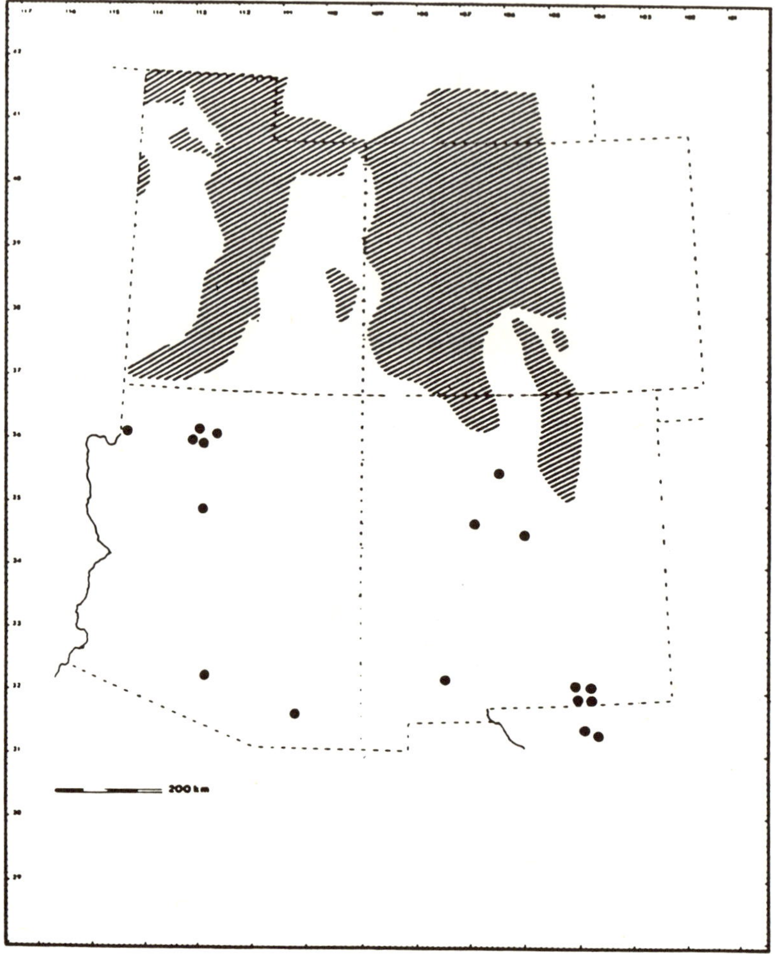

Figure 10.1 Current distribution (stipple) and extralimital fossil sites (dots) of yellow-bellied marmot (*Marmota flaviventris;* figure 10.1a) and montane shrew (*Sorex monticolus;* figure 10.1b) in the American Southwest.

figures 10.1a and 10.1b the modern distributional ranges and extralimital Quaternary sites for the yellow-bellied marmot (*Marmota flaviventris*) and the montane shrew (*Sorex monticolus*), species characteristic of alpine and coniferous forest meadow habitats (Findley 1987). Modern range limits of species only weakly coincide with those attained only a few thousand years ago.

I used correlation analyses to identify interrelationships among dis-

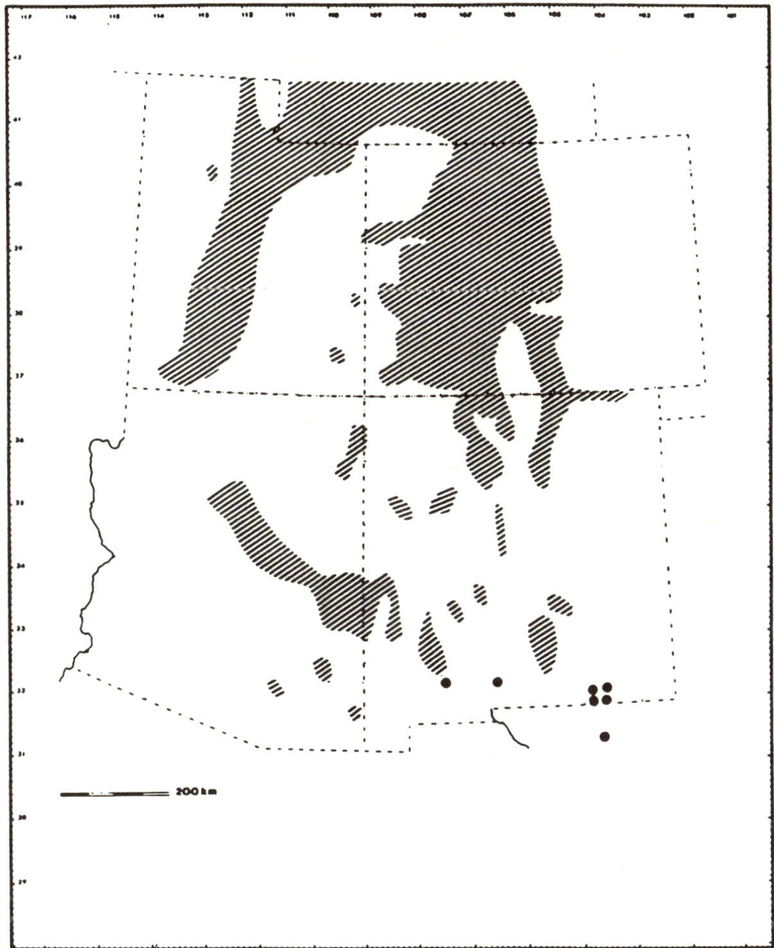

tributional and biological variables (table 10.2). The number of different Southwestern fossil sites and the number of extralimital sites are positively and significantly correlated with the total number of fossil localities from which a species is known. Although these variables are not independent, better fossil records apparently include a greater number of extralimital occurrences during the Pleistocene. Furthermore, representation in the fossil record is positively correlated with the number of vegetational associations that a species occupies ($r = 0.32$), and the latter is correlated with extent of current distribution ($r = 0.30$); however, these correlations are not statistically significant.

Table 10.2 Pearson Product-Moment Correlation Coefficients Among Distributional and Biological Attributes of Boreal Mammal Species

Attribute	Total Fossils	Total Southwestern Fossils	Extralimital Fossils	Log Weight	Range Occurrences
Number of vegetation associations	0.317	0.272	0.281	0.015	0.295
Total fossils		0.914*	0.861*	0.303	0.097
Total Southwestern fossils			0.986*	0.213	0.225
Extralimital fossils				0.195	0.280
Log weight					-0.237

* $p < 0.001$; other coefficients nonsignificant ($p > 0.05$).

Strikingly, variables reflecting the extent of Pleistocene distributions are *not* correlated with current distributional extent. This pattern is true for boreal species as a group, as well as for carnivores, mesic forest herbivores, xeric forest herbivores, and insectivores considered separately. Lack of correlation between the extent of past and present distributions is notable, given positive correlations between (1) fossil representation and number of vegetational associations that a species inhabits, and (2) the latter variable and extent of current distribution. If fossil sites and current distribution were positively correlated, that correlation would lend support to the idea that species that currently have broad distributions were also more broadly distributed during the Pleistocene. This possibility plagues biogeographic analyses that lack reference to a fossil record (Simberloff 1985).

Habitat Affinities and Range Expansions

I next assessed multivariate habitat associations of species using the Jaccard's coefficient and cluster analysis. On the resulting dendrogram (figure 10.2), two main clusters are apparent. One consists of *Sorex cinereus, Spermophilus elegans, Eutamias minimus,* and *Thomomys talpoides,* forms having very broad ecological tolerances, whereas the other comprises the remaining, more selective species. Clusters of species do not generally correspond to the habitat characterizations established by C. Hart Merriam (1890); Vernon Bailey (1931); James Findley and coworkers (1975); or myself (Patterson 1984). In part, such discrepancies may be attributed to the effects of geographic variation in habitat affinities. The associations of Bernard and Brown (1977) reflect distributions throughout 11 western states, whereas the other authors focused on expressly Southwestern distributions. Widespread, polytypic forms having comparatively narrow associations on Southwestern mountain ranges but found elsewhere in sagebrush, alpine meadow, and Great Plains grassland are thus segregated in the dendrogram from more habitat specialists that co-occur with them in the Southwest. Inclusion of data on habitat affinities of boreal species outside the Southwest seems warranted, however, in analyses designed to shed light on the distributions they attained during climatic fluxes of the Pleistocene.

Most segments of the dendrogram have good fossil representation, although some groupings of species are more poorly represented than others (e.g., forest forms, which are loosely clustered in the bottom of the dendrogram). The scanty record for forest species in Southwestern

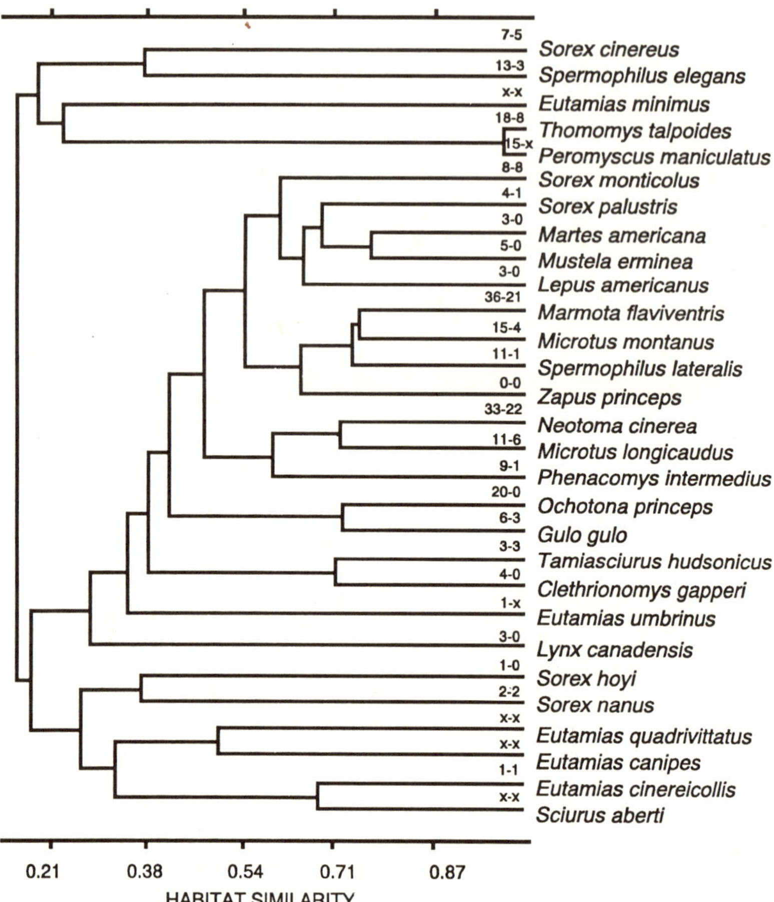

Figure 10.2 Dendrogram of habitat relationships among boreal mammal species in western North America. Similarity was measured by Jaccard's coefficient of association. Numbers before the species name reflect the number of Quaternary fossils in the West and Southwest, respectively; "x" denotes problematic identifications (after Harris 1985).

deposits parallels their poor representation throughout the western United States. This scarcity may signify that forest species were not as widely distributed during the Pleistocene as were forms inhabiting more open vegetative associations, or it may reflect taphonomic biases of different habitats (Harris 1985). Certainly the coniferous forests that these species inhabit were established during the Wisconsin on even small, remote mountain ranges (Van Devender et al. 1979). Forest-dwelling species likely occupied favorable microhabitats in intermontane valleys.

Pleistocene faunas of small isolated ranges document the extent to which boreal species colonized remote and currently inhospitable areas during Pleistocene glacial episodes. Superb Wisconsin records document the Pleistocene fauna of the Guadalupe Mountains in southern New Mexico and Trans-Pecos Texas. Of the 28 mountain ranges I have previously studied (Patterson 1984), which include the Pinaleños, only the Chiricahua Mountains of southeastern Arizona lie farther to the south than the Guadalupes; latitude provides a measure of their isolation from sources for both Pleistocene and Recent colonization. Various cave deposits from the Guadalupe Mountains document Wisconsin populations of 10 of 24 boreal species (42%; table 10.3). These species reflect the full range of habitat affinities among boreal mammal species (figure 10.2). Eight (57%) of the species not recorded from the Guadalupe Mountains lack any fossil record in the Southwest, and their absence in the Guadalupe record is uninformative. Critically, each of the 10 boreal species (100%) known to have occupied the Guadalupe Mountains during the Wisconsin (table 10.3) went locally extinct through natural processes during the last 12,000 years. Four of these species occur today in the nearby Sacramento complex but have not recolonized the Guadalupes through the woodland and steppe vegetation that currently separates these ranges (Küchler 1975).

Other southern localities in the region echo this pattern of glacial colonization and postglacial extinction, albeit with substantially poorer deposits. Harris (1985) reported montane shrew (*Sorex monticolus*), bushy-tailed woodrat (*Neotoma cinerea*), and montane vole (*Microtus montanus*) from pluvial deposits in Anthony Cave, Dona Ana County, New Mexico, at the same latitude as the Guadalupes and 1,555 m in elevation. These species are currently extinct in south-central New Mexico—the nearest existing populations of each lie 160, 410, and 260 km distant, respectively, separated by river valleys or arid basins. Undoubtedly, each occurred widely between these collecting sites during the late Pleistocene. Deposits at Baldy Peak Cave, Luna County, New Mexico (at 1,550 m elevation), record the Pleistocene occurrence of montane shrew, yellow-bellied marmot, and bushy-tailed woodrat (Harris 1985). Although populations of shrews persist on the nearby Mogollon Rim, neither marmots or bushy-tailed woodrats remain within 400 km of this site today.

Table 10.3 Quaternary Records of Boreal Mammals on the Guadalupe Mountains

Boreal Species (ranked by range occupancy)	On Guadalupe Mts. During Wisconsin	Approx. Distance to Nearest Surviving Population (km)
Sorex monticolus	X	125[b]
Microtus longicaudus	X	125[b]
Tamiasciurus hudsonicus	X	125[b]
Eutamias minimus	—[a]	175[b]
Spermophilus lateralis	—	310
Thomomys talpoides complex	X	430
Neotoma cinerea	X	440
Sorex palustris	X	450
Clethrionomys gapperi	—[a]	320
Marmota flaviventris	X	450
Sorex nanus	X	175
Zapus princeps	—[a]	450
Microtus montanus	—	400
Mustela erminea	—	400
Ochotona princeps	—	450
Sorex cinereus	X	490
Lepus americanus	—[a]	440
Martes americana	—	425
Phenacomys intermedius	—	445
Gulo gulo	—[a]	690
Lynx canadensis	—[a]	690
Sorex hoyi	—[a]	985
Eutamias umbrinus	—[a]	
Spermophilus elegans	X	790

Note: In the context of all boreal mammal species now occurring in the Southwest. None of these species now occurs on the Guadalupes, demonstrating the extent of glacial colonizations, the ravages of postglacial extinction, and the impotence of postglacial colonization in the biogeography of this range.

[a] Species without Wisconsin records in the Southwest.

[b] Populations lying along coniferous woodland dispersal corridor (Sullivan 1985) with narrow shrub-steppe gap immediately north of Guadalupes.

Pleistocene Range Dynamics

An extensive fossil record for the southwestern United States demonstrates that many boreal mammals attained greatly expanded geographical ranges during the Pleistocene. There can be no question that during "glaciopluvial" periods (glacial episodes of greater effective moisture), boreal species colonized distant, isolated mountain ranges, in many cases also occupying the intervening lowlands. The same record shows that during the Holocene all lowland populations and many montane populations of boreal mammals have gone extinct due to natural causes (Lundelius et al. 1983). These findings substantiate and quantify earlier observations that nearly half of the Pleistocene faunas in the Southwest comprised taxa now extinct or extralimital (Schultz and Howard 1935; Ayer 1936); they also corroborate and complement Donald Grayson's (1987) recent studies on postglacial extinctions of boreal mammals in the Great Basin, as well as Harris's (1990) analyses of Pleistocene distribution limits in the Southwest. Grayson's analyses surpass those presented here in using stratified and dated samples to establish precise sequences of colonization-extinction events. In turn, the superior fossil record for the Southwest and its richer boreal fauna permit inferences covering more of the boreal biota and many more study sites (Harris 1990).

The extent of Pleistocene geographic ranges of boreal species appears to be uncorrelated to current distributional ranges. Biologists would expect to find a correlation between past and present ranges if fossil distributions were simply tattered samples of modern ones and both had been affected by the same factors. The lack of correlation documented for Southwestern boreal mammals, but as yet known for no others, is crucial to studies of historical derivation based on modern distribution patterns. Biologists cannot attribute modern distributional patterns to any historical process, either to extinction or to colonization, without first accurately documenting the initial species pool (Simberloff 1985).

The fossil record documents the utter disappearance of boreal mammals from the Guadalupe Mountains via local extinctions. As discussed above, all 10 species known from Wisconsin fossils went locally extinct during the postglacial period. However, populations of 4 of these species still occur today in the nearby Sacramento Mountains, which are connected to the Guadalupes by continuous forest and woodland corridors. This situation seems inconsistent with the view that woodlands constitute leaky barriers to montane mammals (Davis et al. 1988)—were this true, surviving populations in the Sacramento Mountains could readily recolonize the Guadalupes once populations in the latter went locally extinct. It is interesting to note that 3 of the 4 species also show significant "isolation effects" (correlation of distribution with distance from a source area). Some biologists (e.g., Lomolino et al. 1989) have interpreted significant isolation effects as evidence of a modern-day dispersal ability; however, no boreal species absent from the Guadalupes during the Pleistocene is known to have colonized the range during postglacial times.

Relevance to the Issue of Distributional Dynamics

Some researchers believe that biogeographic patterns shown by mammals in the southern Rocky Mountains are more dynamic than those of the Great Basin (Davis et al. 1988; Lomolino et al. 1989). Only the mammals of the southern Rocky Mountains show historical range expansions and statistically significant isolation effects. Yet fossil data in both regions present overwhelming evidence for postglacial extinctions. How dynamic are these montane distributions?

Circumstantial evidence for inter-range movements of certain montane mammal species is strong. Species living on Southwestern mountaintops are more diverse and heterogeneous than Great Basin species with respect to faunal origins. As one indication of diversity, no two faunal lists in studies of Southwestern mammals are identical, underscoring the subjectivity of defining a reference species pool. Further, elevational zonations of habitats in the Southwest are complex and geographically variable. Undoubtedly, habitats that are fully isolated from one another at the southern end of this biotic province are broadly continuous at the northern end, and are thus more apt to exchange faunal elements there.

Yet mountain ranges in the Southwest and their mammal faunas seem far more isolated than nonisolated "mainland" areas. One of my colleagues and I (Patterson and Atmar 1986) compared the species-area regression slope for 28 isolated Southwestern ranges with the slope for an equivalent range of nonisolated areas within the main block of the southern Rocky Mountains using the approach developed by James Brown (1971) to assess the effects of isolation on species-richness patterns (Lawlor 1986). The resulting slopes, 0.36 for isolated ranges versus 0.07 for nonisolated ranges, differ significantly. Because the species pools and habitats are similar in both regressions, we presumed that the higher slope for isolated ranges reflects the importance of extinction, unopposed by recurrent colonization.

Apparently, the isolation effects demonstrated by other researchers (e.g., Davis et al. 1988; Lomolino et al. 1989) using different species pools represent supplemental patterns that do not contradict the prevailing effect of area. To evaluate the relative correlations of these variables to species richness patterns, I employed a stepwise regression model using latitude, longitude, elevation, logarithm of mesic forest area, logarithm of forest area, isolation, and isolation squared as independent variables (Patterson 1984; Davis et al. 1988). A maximum-R^2 procedure on logarithm of mammal species yields the following: the logarithm of mesic area enters the regression equation first ($R^2 = 0.74$), isolation second (improved $R^2 = 0.08$), elevation third (improved $R^2 = 0.03$), and longitude fourth (improved $R^2 = 0.02$). Each term in the expanded model is significant ($p < 0.05$). Under island biogeographic theory, the regression terms for area, isolation, and elevation may be interpreted as reflecting extinction, colonization, and habitat diversity, respectively. Regardless of interpretation, area effects dominate all others.

Researchers in the past have failed to recognize that the isolation effects which are implicated by regression analyses lack any temporal information. Whereas biologists may usefully attribute isolation effects to differential colonization under the MacArthur-Wilson paradigm, the existence of isolation effects does not identify the timing of such events. This element looms as a significant complication in interpretation because Pleistocene range expansions by boreal species were unequal in the Southwest. Analyses of fossils show that mountains in southern Arizona and in southwestern New Mexico were never joined to northern forests by continuous forested habitat, even during the height of the Wisconsin (Harris 1990; Lomolino et al. 1989:Fig. 2), but were instead isolated throughout the Pleistocene by a broad "sagebrush steppe-woodland" zone similar to the "sea" surrounding most Great Basin mountain ranges (Brown 1971). Pleistocene colonization of these ranges was probably limited to adept dispersers and habitat generalists. Interestingly, ranges in the southern Great Basin were probably isolated in a similar fashion (Wells 1983). If Pleistocene colonization of these ranges was incomplete, then modern-day richness and composition patterns should show isolation effects even in the absence of any ongoing dispersal. Fully a third of the mountain ranges studied by M. V. Lomolino, James Brown, and Russell Davis (1989) were historically isolated from the main expanse of Rocky Mountain biotas, whereas only 3 of 28 I analyzed in 1984 were so isolated.

Range Expansion by Species of Austral and Boreal Origins

Statistics aside, several montane mammals demonstrably move through suitable habitats between mountain ranges and thereby colonize new ranges. The range of Mexican voles (*Microtus mexicanus*), for example, appears to be extending northward in New Mexico (Findley et al. 1975), based on twentieth-century collecting records (see also Davis and Callahan 1992). Similar evidence indicates that yellow-nosed cotton rats (*Sigmodon ochrognathus*) have recently crossed intermontane valleys in southeastern Arizona (Davis and Dunford 1987) and that Abert's squirrels (*Sciurus aberti*) are now expanding in western Colorado (Davis and Brown 1989). Each of these instances refutes the hypothesis that all range expansions occurred during the Pleistocene and that post-Pleistocene range changes have been strictly attritional.

However, two points make these observations generally irrelevant to boreal biogeography. First and most important, all of these species are austral in origin. That is, each has a center of origin in the Sierra Madre Occidental and adjacent areas of Mexico. The same is true for *each* of the 19 species that Davis and Callahan (1992) found were expanding their ranges northward in the Southwest. Second, none of the species under discussion is recorded from the southern Rocky Mountains region during the Wisconsin glaciation. Each has apparently expanded its range from Mexican refugia during late Wisconsin or early postglacial times, at a time when boreal taxa were undergoing range contractions (Harris 1990). These case studies of dispersal do not serve as evidence of boreal range dynamics; rather, they demonstrate that the distribution patterns of austral species have generally been complementary to, not similar to, those of boreal species. During glacial episodes, the distributions of austral species contracted in the Southwest while boreal species were undergoing range expansions; with warmer, drier Holocene climates, boreal ranges shrank while conditions ameliorated for austral taxa (Patterson 1984).

Distinctions between faunas may also be apparent in the incidence patterns presented by Lomolino, Brown, and Davis (1989). Boreal species appear to show mainly area effects, as expected by the postglacial extinction hypothesis — both species that showed only area effects were boreal in origin. By contrast, the two species that exhibited only isolation effects included an austral species and an indigenous species of uncertain relationships, species thought to have expanded their ranges into the Southwest during the postglacial period. It would be interesting to examine other montane mammals to see whether this dichotomy between species of austral and boreal origins broadly holds.

These observations and others suggest a complex Pleistocene history for mammals that live in Southwestern mountains. On the one hand, detailed historical records for austral species show them to have highly dynamic ranges; the range limits of these species may indeed approximate a dynamic equilibrium. On the other hand, most boreal species became fragmented throughout the southern Rocky Mountains by Recent climate changes, and since that time, local extinction has been the sole documented agent for range changes. Undoubtedly, each group of species will exhibit variation, but the contrasting patterns shown by these groups ought to be mutually illuminating. Obviously, to derive clear and homogeneous patterns of biogeographic distribution, researchers must determine centers of origin and homogeneous geographic theaters independently and a priori. Testing these distributional hypotheses using population genetic theory and markers represents a logical "next step." Analyses in progress by Robert Sullivan, David Hafner, and Timothy Lawlor promise such insights.

Relevance to the Fauna of the Pinaleño Mountains

The implications of these findings for the fauna of the Pinaleños seem straightforward. Strong evidence indicates that populations of yellow-nosed cotton rats and Abert's squirrels, as well as numerous woodland species, are either recent invaders, introductions (as with Abert's squirrels in the Pinaleños), or continuously distributed through the woodlands of southeastern Arizona (Hoffmeister 1986; Davis and Dunford 1987; Davis and Callahan 1992). Most of these species were absent from Southwestern mountain ranges during the height of the Pleistocene glaciations and found refuge from cold, wet climates farther south in Mexico. Modern Southwestern populations of such species should be weakly if at all differentiated from conspecifics elsewhere. Furthermore, their populations should communicate—through occasional dispersal and gene flow—with other populations on other mountain ranges in the region. Both the demographics and the genetics of these populations should follow an isolation-by-distance model (Wright 1978).

However, the boreal taxa that today occur in the Pinaleños are

Table 10.4 Current Distributions of Montane Mammals in the Southwest

Range	Area of All Montane Habitats (km^2)	Montane Mammal Species*																									
Southern Rocky Mts.	58,508	A	B	C	D	E	F	G	H	I	J	K	L	M	N	O	P	Q	R	S	T	U	V	W	X	Y	Z
San Juan	33,120	A	B	C	D	E	F	G	H	I	J	K	L	M	N	O	P	Q	R	S	T	U	V	W	X		
Sangre de Cristo	19,604	A	B	C	D	E	F	G	H	I	J	K	L	M	N	O		Q	R	S	T	U	V	W	X		
Pike's Peak	17,169	A	B	C	D	E	F	G	H	I	J	K	L	M	N	O	P	Q	R	S	T	U					
Jemez	3,288	A	B	C	D	E	F	G	H	I	J	K	L	M		O	P	Q	R	S	T						
Uncompaghre Plateau	2,953	A	B	C	D	E		G	H	I	J	K			N	O	P										
LaSal	791	A	B	C	D	E		G		I	J	K			N	O		Q					V				
Rabbit Hills	2,161	A	B	C	D	E			H	I	J	K	L		N	O											
White	18,630	A	B	C	D	E	F	G	H			K	L				P										
Abajo	1,826	A	B	C	D	E	F	G		I	J	K															
Chuska	3,257	A	B	C	D	E	F	G	H	I	J																
Sandia	183	A	B	C	D	E	F	G						M					R								
Mogollon	3,531	A	B	C	D	E	F		H				L				P										
Tavaputs Plateau	639	A	B	C		E				I	J				N												
Mt Taylor	852	A	B	C	D	E	F			I																	

Manzano	791	A	B	C	D	E	F							M
Sacramento	3,531	A	B	C	D	E		G						M
Black Range	1,583	A	B	C	D	E	F		H					
San Francisco	18,630	A	B	C	D	E	F		H					
Zuni	2,892	A	B	C	D	E	F							
San Mateo	670	A	B	C	D		F						L	
Capitán	487	A	B	C	D	E								
Magdalena	183	A	B	C			F						L	
Pinaleño	244	A	B		D	E								
Santa Catalina	122	A	B				F							
Chiricahua	122	A	B											
Organ	50			C										
Guadalupe	244			C										

Source: Patterson and Atmar 1986.

Note: Ranked by the species richness of mountain ranges and range occurrences of species, these faunas constitute a distinctly nested series, the species in smaller faunas being a subset of those in larger ones. Rankings of species by range occurrences also indicate their tendency to be prone to postglacial extinction.

* Key to montane mammals: A, *Sorex monticolus*; B, *Peromyscus maniculatus* (deer mouse; eurychore); C, *Eutamias quadrivittatus* group (Colorado chipmunk group; autochthon); D, *Tamiasciurus hudsonicus*; E, *Microtus longicaudus*; F, *Sciurus aberti* (Abert's squirrel; austral); G, *Eutamias minimus*; H, *Spermophilus lateralis*; I, *Thomomys talpoides* complex; J, *Neotoma cinerea*; K, *Sorex palustris*; L, *Clethrionomys gapperi*; M, *Sorex nanus*; N, *Marmota flaviventris*; O, *Zapus princeps*; P, *Microtus montanus*; Q, *Ochotona princeps*; R, *Mustela erminea*; S, *Sorex cinereus*; T, *Lepus americanus*; U, *Phenacomys intermedius*; V, *Martes americana*; W, *Gulo gulo*; X, *Lynx canadensis*; Y, *Sorex hoyi*; Z, *Spermophilus elegans*.

merely a subset of those that colonized Southwestern mountain ranges during the Pleistocene. Many more boreal species colonized the Guadalupe Mountains (table 10.3), which have less forest area and lower peak elevations and are even more distant from colonization sources. However, all glacial colonists of the Guadalupes, and many of those on other small isolated mountain ranges, have become extinct during postglacial times. No evidence—anecdotal, statistical, or otherwise—suggests that any boreal mammals have naturally recolonized the Pinaleños (or any other isolated mountain range in the Southwest) during Recent times. The modern boreal fauna of the Pinaleños appears strictly attritional—populations of the red squirrel and other boreal species being fragmented relicts of more widespread Pleistocene distributions. Throughout the time of their isolation, these populations may have accumulated genetic distinctions through mutations, adaptation, and genetic drift, thus becoming genetically differentiated from populations elsewhere. Given the variety of these agencies, and their variable impacts, divergence patterns should be rather idiosyncratic, as for example the correlation of forested area with heterozygosity in the least chipmunk (Sullivan 1985), reflecting divergence owing to genetic drift and effective population size (Chapter 12).

The susceptibility of individual species of mammals to local extinction is a predictable and apparently deterministic consequence of body size, trophic position, and habitat specialization (Brown 1971; Patterson 1984). Extinction risk appears highest in large carnivores (especially among habitat specialists) and lowest in small herbivores (especially habitat generalists). Because traits affecting extinction risk do not vary significantly among mountain ranges and most ranges began the Holocene with similar faunas, the same species should have gone extinct in approximately the same order on each of the mountain ranges. As Wirt Atmar and I (Patterson and Atmar 1986) have argued, this logical conclusion accounts for the "nested subset" character of Southwestern mountain ranges, where each mountain range supports a subset of the boreal mammal species found in richer faunas (table 10.4).

As Atmar and I have developed more fully elsewhere (Atmar and Patterson 1993), a "nested" distribution matrix like that shown in table 10.4 can be viewed from two perspectives. From a faunal standpoint (top to bottom), faunas are ranked in terms of the extent to which they have been winnowed by local extinction. Because extinction rates are closely tied to area, faunas ranked by species richness are also generally ranked by area (the largest [on the top] experiencing the lowest cumulative extinctions, the smallest [on the bottom] ravaged by local extinction). From a species standpoint (left to right), species are ranked by their extinction risk (the leftmost species finding even the smallest mountain ranges hospitable, whereas the rightmost species is the first to disappear when ranges become isolated). When both rankings are taken into account simultaneously, different species are at different risk on different mountain ranges. Under the nested subset model, the diagonal from lower left to upper right defines a set of populations most at risk of extinction.

The Pinaleño Mountains support only two montane mammals that are not also found in the smaller, adjacent Chiricahuas: the red squirrel (D) and the long-tailed vole (E). Both populations are candidates for extinction under the nested subset theory. In fact, both were placed "under Notice of Review" by the U.S. Fish and Wildlife Service for listing as threatened or endangered (*Federal Register* 30 December 1982), with the red squirrel subsequently listed as endangered. The "nested subset" diagonal such as that in table 10.4 serves to identify vulnerable populations on *each* mountain range in the Southwest, doing so in the context of their regional distributions. This approach offers a simple, inexpensive, yet highly reliable approach to identifying "minimum viable population size" and other quantities of interest to conservationists and wildlife managers.

Acknowledgments

Aspects of this study were first presented at a 1987 symposium of the American Society of Mammalogists entitled "Historical Zoogeography of the Southwest." I thank Dave Hafner and Bob Sullivan for suggesting that presentation and Bob Hoffmann for inviting me to relate those findings to the Mt. Graham workshop. The stimulation afforded by countless biogeographic discussions and correspondence with W. Atmar, J. H. Brown, R. Davis, J. M. Diamond, B. A. Harney, T. E. Lawlor, M. V. Lomolino, W. T. Stanley, and R. M. Sullivan is greatly appreciated. M. A. Rogers and L. D. Brady assisted this overview in a variety of ways, and C. L. Friedman drafted figure 10.1. The study would have been impossible without the exhaustive compilations and syntheses of D. M. Armstrong, S. D. Durrant, J. S. Findley and colleagues, A. H. Harris, and D. F. Hoffmeister. I thank P. V. Wells, T. M. Van Devender, and two anonymous reviewers for criticisms of earlier drafts.

References

Armstrong, D. M. 1972. Distribution of mammals in Colorado. *Monographs of the Museum of Natural History* [The University of Kansas] 3:x, 1–415.

Atmar, W., and B. D. Patterson. 1993. The measure of order and disorder in the distribution of species in fragmented habitat. *Oecologia* 96:373–82.

Ayer, M. Y. 1936. The archeological and faunal material from Williams Cave, Guadalupe Mountains, Texas. *Proceedings of the Academy of the Natural Sciences* [Philadelphia] 88:599–618.

Bailey, V. 1931. Mammals of New Mexico. *North American Fauna* 53:1–412.

Bernard, S. R., and K. F. Brown. 1977. *Distribution of Mammals, Reptiles, and Amphibians by BLM Physiographic Regions and A. W. Küchler's Associations for the Eleven Western States.* Technical Note 301. U.S. Department of the Interior, Bureau of Land Management.

Brown, J. H. 1971. Mammals on mountaintops: Nonequilibrium insular biogeography. *American Naturalist* 105:467–78.

Brown, J. H. 1978. The theory of insular biogeography and the distribution of boreal birds and mammals. *Great Basin Naturalist Memoirs* 2:209–27.

Davis, R., and D. E. Brown. 1989. Role of post-Pleistocene dispersal in determining the modern distribution of Abert's squirrel. *Great Basin Naturalist* 49:425–34.

Davis, R., and J. R. Callahan. 1992. Post-Pleistocene dispersal in the Mexican vole (*Microtus mexicanus*): An example of an apparent trend in the distribution of Southwestern mammals. *Great Basin Naturalist* 52:262–68.

Davis, R., and C. Dunford. 1987. An example of contemporary colonization of montane islands by small, nonflying mammals in the American Southwest. *American Naturalist* 129:398–406.

Davis, R., C. Dunford, and M. V. Lomolino. 1988. Montane mammals of the American Southwest: The possible influence of post-Pleistocene colonization. *Journal of Biogeography* 15:841–48.

Diamond, J. M. 1984. "Normal" extinctions of isolated populations. In *Extinctions,* ed. M. H. Nitecki, pp. 191–246. University of Chicago Press, Chicago.

Findley, J. S. 1969. Biogeography of Southwestern boreal and desert mammals. In *Contributions in Mammalogy,* ed. J. K. Jones, pp. 113–28. Miscellaneous Publications of the Museum of Natural History 51. University of Kansas, Lawrence.

Findley, J. S. 1987. *The Natural History of New Mexican Mammals.* New Mexico Natural History Series. University of New Mexico Press, Albuquerque.

Findley, J. S., A. H. Harris, D. E. Wilson, and C. Jones. 1975. *Mammals of New Mexico.* University of New Mexico Press, Albuquerque.

Grayson, D. K. 1987. The biogeographic history of small mammals in the Great Basin: Observations on the last 20,000 years. *Journal of Mammalogy* 68:359–75.

Grinnell, J., and H. S. Swarth. 1913. An account of the birds and mammals of the San Jacinto area of southern California [with remarks upon the behavior of geographic races on the margins of their habitats]. *University of California Publications in Zoology* 10(10):197–406.

Harris, A. H. 1985. *Late Pleistocene Vertebrate Paleoecology of the West.* University of Texas Press, Austin.

Harris, A. H. 1990. Fossil evidence bearing on southwestern mammalian biogeography. *Journal of Mammalogy* 71:219–29.

Hoffmeister, D. F. 1986. *Mammals of Arizona.* University of Arizona Press, Tucson, and Arizona Game and Fish Department, Phoenix.

Küchler, A. W. 1975. *The Potential Natural Vegetation of the Conterminous United States, 1:3,168,000.* Special Publications of the American Geographical Society.

Lawlor, T. E. 1986. Comparative biogeography of mammals on islands. In *Island Biogeography of Mammals,* ed. L. R. Heaney and B. D. Patterson, pp. 99–125. Linnean Society of London and Academic Press, London.

Lomolino, M. V., J. H. Brown, and R. Davis. 1989. Island biogeography of montane forest mammals in the American Southwest. *Ecology* 70:180–94.

Lundelius, E. L., Jr., R. W. Graham, E. Anderson, J. Guilday, J. A. Holman, D. W. Steadman, and S. D. Webb. 1983. Terrestrial vertebrate faunas. In *Late Quaternary Environments of the United States,* vol. 1, *The Late Pleistocene,* ed. S. C. Porter, pp. 311–53. University of Minnesota Press, Minneapolis.

MacArthur, R. H., and E. O. Wilson. 1967. *The Theory of Island Biogeography.* Monographs in Population Biology 1. Princeton University Press, Princeton.

Merriam, C. H. 1890. Results of a biological survey of the San Francisco Mountain region and desert of the Little Colorado in Arizona. *North American Fauna* 3:1–136.

Metcalf, A. L. 1977. Some Quaternary molluscan faunas from the northern Chihuahuan Desert and their paleoecological implications. In *Transactions of the Symposium on the Biological Resources of the Chihuahuan Desert,* ed. R. H. Wauer and D. H. Riskind, pp. 53–66. National Park Service, Transactions and Proceedings Series Number 3. GPO, Washington, D.C.

Patterson, B. D. 1980. Montane mammalian biogeography in New Mexico. *Southwestern Naturalist* 25:33–40.

Patterson, B. D. 1984. Mammalian extinction and biogeography in the southern Rocky Mountains. In *Extinctions,* ed. M. H. Nitecki, pp. 247–93. University of Chicago Press, Chicago.

Patterson, B. D., and W. Atmar. 1986. Nested subsets and the structure of mammalian faunas and archipelagos. In *Island Biogeography of Mammals,* ed. L. R. Heaney and B. D. Patterson, pp. 65–82. Linnean Society of London and Academic Press, London.

Schultz, C. B., and E. B. Howard. 1935. The fauna of Burnet Cave, Guadalupe Mountains, New Mexico. *Proceedings of the Academy of the Natural Sciences* [Philadelphia] 87:273–98.

Simberloff, D. S. 1985. Extinctions [Review]. *Auk* 102:429–31.

Smith, G. R. 1978. Biogeography of intermountain fishes. *Great Basin Naturalist Memoirs* 2:17–42.

Sneath, P. H. A., and R. R. Sokal. 1973. *Numerical taxonomy.* W. H. Freeman and Co., San Francisco.

Spaulding, W. G., E. B. Leopold, and T. R. Van Devender. 1983. Late Wisconsin paleoecology of the American southwest. In *Late Quaternary Environments of the United States,* vol. 1, *The late Pleistocene,* ed. S. C. Porter, pp. 259–93. University of Minnesota Press, Minneapolis.

Sullivan, R. M. 1985. Phyletic, biogeographic, and ecologic relationships among montane populations of least chipmunks (*Eutamias minimus*) in the Southwest. *Systematic Zoology* 34:419–48.

Van Devender, T. R., and W. G. Spaulding. 1979. Development of vegetation and climate in the southwestern United States. *Science* 204:701–10.

Van Devender, T. R., W. G. Spaulding, and A. M. Phillips III. 1979. Late Pleistocene plant communities in the Guadalupe Mountains, Culberson County, Texas. In *Biological Investigations in the Guadalupe Mountains National Park, Texas,* ed. H. H. Genoways and R. J. Baker, pp. 13–30. National Park Service, Transactions and Proceedings Series Number 4. GPO, Washington, D.C.

Wells, P. V. 1983. Paleobiogeography of montane islands in the Great Basin since the last glaciopluvial. *Ecological Monographs* 53:341–82.

Wright, S. 1978. *Evolution and the Genetics of Populations,* vol. 4, *Variability Within and Among Populations.* University of Chicago Press, Chicago.

Vulnerability to Extinction of Small Isolated Populations

Biology and the Red Squirrel

You may easily know this little workman by his chips. On sunny hillsides around the principal trees they lie in big piles—bushels and basketfuls of them, all fresh and clean, making the most beautiful kitchen-middens imaginable.

JOHN MUIR, *The Wilderness World of John Muir*

The first chapter of Part 5 provides an overview of the vulnerability to extinction of small isolated populations, such as the Mt. Graham red squirrel, and sets the stage for the remaining three chapters, which deal more specifically with different aspects of the biology of red squirrels. The first chapter self-consciously provides a thoughtful transition, a logical bridge, between the previous part and this one. At the close of his overview the author, Lawrence Heaney, discusses the legal roles and ethical responsibilities of administrators of the University of Arizona.

The second chapter, by Robert Sullivan and Terry Yates, presents an analysis of the genetic variation and systematics of red squirrels in the western United States, including data for the Pinaleños population, and the relevance of the genetic differentiation and structuring of these populations to conservation biology.

In the third chapter Christopher Smith provides a comprehensive treatment of the comparative ecology of red squirrels and other terrestrial and arboreal squirrels. Such knowledge is necessary to understand and manage both populations and the specific associations and interactions of these species in natural communities.

The fourth chapter, by Paul Young, analyzes results from the monitoring program for the red squirrel population in the Pinaleño mountains, as required under agreements pertaining to the construction of the Mt. Graham International Observatory.

CHAPTER ELEVEN

Population Vulnerability of Mammals in Isolated Habitats

Lawrence R. Heaney

The spotlight of this volume is on the Mt. Graham red squirrel, and the current status and prospects for the biota of the Pinaleño Mountains on which it lives, as in the 1989 workshop on which this volume is based. However, as some of the other chapters indicate, the issues of greatest importance for this one species pertain equally strongly to isolated populations and habitats worldwide, especially as natural habitats become increasingly reduced in area and fragmented by human activity. The ubiquity of these issues has resulted in repeated studies and tests that help to provide a clear context by which to gauge the status of the Mt. Graham red squirrels. This chapter provides an overview of that context.

For the last 15 years, in addition to studying the ecology of tree squirrels (Heaney 1984b), I have studied the ecology and evolution of mammals in island archipelagos, with a special interest in the origin and maintenance of patterns of biological diversity (Heaney 1984a, 1986, 1991b, 1993; Heaney and Rickart 1990), and it is on that basis that I hope to make a contribution to understanding the biota of the Pinaleño Mountains. My research has focused on quantitative documentation of the processes that shape diversity patterns — namely, colonization, speciation, and extinction. I have done all of this work in Southeast Asia, especially the Philippines (where I have been able to put this information to use in helping the Philippine government design a national park system that we hope will protect the greatest possible portion of the Filipino natural heritage; Heaney 1993). My role in this volume, then, is not so much to provide new information on red squirrels or the Pinaleños as to provide background information on the conservation biology of mammals and to ask some hard questions about past and current activities in the Pinaleños.

At the outset, I wish to emphasize two observations. First, studies of diversity and extinction are not new. It was, in fact, observations of diverse but restricted faunas and the fossil remains of extinct animals that led Charles Darwin and Alfred Russel Wallace to develop independently the theory of evolution by natural selection more than 135 years ago. Biologists and resource managers should not overlook the importance of the existing broad body of information on ecological and evolutionary biogeography as a source of context and perspective for investigating individual cases.

Second, several authors of chapters in this volume have successfully adopted the perspective of preserving a unique community of organisms, rather than a single species. I applaud this approach wholeheartedly. To be successful in conservation on a national and international basis, we must shift our emphasis from protecting single species to protecting habitats and communities. Only in this way can we make the critical step from recovery of endangered species to prevention of endangerment. I believe that the latter must be made the primary goal of conservation if we are to have any hope of retaining our nation's and planet's natural diversity. Unfortunately, no other legal mechanism for the protection of endangered communities approaches the effectiveness of the Endangered Species Act. We should recognize that the entire ecosystem of the high-elevation Pinaleños, with its dozen or more unique organisms, is endangered because of the Mt. Graham observatory project; but the legal recognition of a single species makes that one species the necessary legal focus of the current debate. I do not question the importance of that one species, but I emphasize the presence of other equally vulnerable and equally unique organisms in the Pinaleños. All of these creatures make it a special place for research biologists and conservationists alike.

Assessing Population and Community Vulnerability

In this essay, I approach the daunting task of assessing the current status of the Pinaleños and its red squirrels by using an analogy. Several authors in this volume have used dramatic imagery by referring to the Pinaleños and the other high peaks of the Southwest as "sky islands." Although the expression is dramatic, it is not mere imagery; the high-elevation communities are indeed composed principally of species that are cut off by many kilometers—often hundreds of kilometers—from other populations of their species. They are surrounded by habitat that is as uninhabitable for them as is seawater for animals on the oceanic islands of Southeast Asia. The analogy that I use here (following its use by many others) is to view these sky islands as if they were actual islands and to borrow liberally from the extensive research on island biogeography begun by Darwin and his contemporaries.

Conservation biologists currently view island biogeography as a powerful tool because natural circumstances have given us the opportunity to study the effects of extinction on islands all over the globe. This situation has occurred principally because of the rise of sea level by 120 m (400 feet) as continental glaciers melted at the close of the "ice ages," ending about 8,000 years ago, which isolated many former continental land areas as islands (Fairbanks 1989). Cut off from other populations and beset by changing climate and a vastly declining habitat area, the isolated plant and animal populations underwent reduction and often extinction (Heaney 1984a, 1986, 1991a). In the Southwest, warming and drying conditions gradually eliminated forest habitats in the lowlands, leaving only some patches on the high peaks (Chapters 8–10, 12). The isolated populations of the Pinaleños are now beset a second time by decreasing habitat area and perhaps changing climate as well, this time induced by humans (see Chapter 15).

For many years, biologists viewed the islands of Southeast Asia as the premier example of island biogeographic studies. Now, however, the vertebrates living on the mountaintops of the Southwest stand as the outstanding case study of island biogeography: data on distribution, ecology, and fossil history are superb, unparalleled even in Southeast Asia where the claim of historical primacy can be made (Chapter 8). The montane Southwest is a vibrant, dynamic case study area, with new data, new questions, and new insights contributing yearly to our understanding of extinction and evolution. Current research in the Pinaleños is certain to contribute to this effort.

By studying these processes at many sites around the globe, researchers have clearly documented a series of general patterns and are investigating others. All of the patterns are generalizations that are essentially probabilistic: all show variation—no science, not even astrophysics, can make absolute predictions for individual events. But the strength of the patterns is great, and, most important, biologists clearly have the

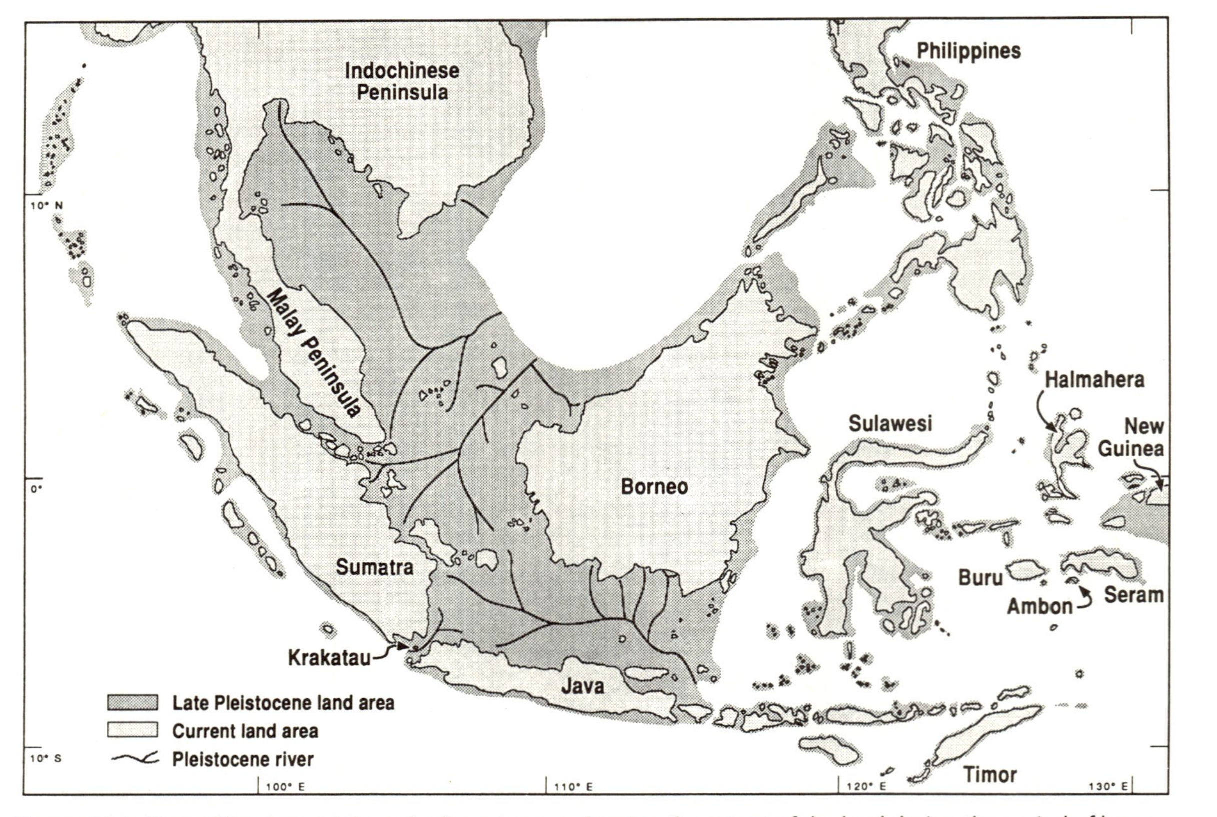

Figure 11.1 Map of Southeast Asia and adjacent areas, showing the extent of dry land during the period of low sea level that marked the final phase of the Pleistocene glaciations (from Heaney 1991b).

ability to identify vulnerable populations on the basis of these patterns. I have proposed the seven following questions as a means of presenting these patterns for discussion.

1. Can areas of high biological diversity and uniqueness be identified on the basis of current information?

The answer to this question seems completely clear; yes, these areas can and, in many instances, have been identified. As Russell Davis explains (Chapter 8), each of the historically defined regions in a given area has a unique fauna. In this case, the distinctive montane faunas are in the Sierra Madres and the Rocky Mountains, with the Pinaleños lying between and representing a rare area of faunal intermixing (Chapters 9 and 10). The comparable case in Old World biogeography involves Asia and Australia as centers of endemism, with the island of Sulawesi (also known as Celebes) as the classical example of a uniquely intermixed fauna (figure 11.1; Musser 1988). Davis's comparison between the Pinaleños and Sulawesi is an unusually insightful analogy (Chapter 8); I rather wish I had thought of it first.

We are able to identify these faunas primarily because they share common histories. In the case of Australasia, the Asian fauna that is now scattered on islands was once continuous because sea level was low enough that the islands were united as part of the mainland; likewise, the islands with Australian faunas, including New Guinea, were united with the Australian continent (Heaney 1991a). Only islands scattered between the continental shelves have been continuously isolated, and only they have intermixed faunas. In the case of the Southwest, the high-elevation mammalian faunas of the Sierra Madres and the Rockies have also remained isolated, with only some species having broad ecological tolerance moving between them during the cool and moist period of the late Pleistocene; only isolated peaks such as those of the Pinaleños received migrants from both mountain massifs (Chapters 9 and 10).

I should point out that mammals are not unique in showing such patterns; essentially any taxon that has comparably low dispersal ability and high habitat affiliation should be similar. In other words, if these patterns exist within the mammalian community, the plant, insect, and snail communities will very likely exhibit strongly (but not perfectly) parallel unique features. This concordance of pattern means that biologists can take the well-known taxa and use them to define necessary conservation actions, with the knowledge that those actions will also simultaneously protect many of the other taxa that are more poorly known.

2. Can patterns of diversity be predicted on the basis of current data?

In addition to defining patterns of distribution, biologists and resource managers must also consider diversity, defined here as the number of species in a given area. Studies on islands in nearly every portion of the globe have shown that diversity of mammals (and most other groups) is very strongly associated with the area of the island they occupy (Chapter 10); the smaller the island, the fewer the species. This pattern is so regular and pervasive that exceptions are publishable. Research has shown that this pattern holds as well for the mammals on the high-elevation habitat islands in the Southwest (Brown 1971, 1978; Patterson 1984; Patterson and Atmar 1986; Lomolino et al. 1989). At the species level, this pattern occurs primarily because a reduction in area of habitat nearly always causes a reduction in population size, which increases the vulnerability of the population to the vagaries of climate, to inbreeding depression, to disease, and probably to other negative factors as well. At the community level, the reduction in species probably is due in part to the general correlation between island area and habitat diversity: a small mountain may not have a marshy meadow, for example, whereas a large mountain almost certainly will. We are thus using area as an easily measured surrogate for two more complex, less easily measured factors: population size and habitat diversity. The near-ubiquity of the diversity-area correlation serves as abundant justification for continuing to use area as our measure.

In the problem at hand, the correlation between diversity and area has a simple and direct implication for the Pinaleños—a reduction in a given habitat patch (of spruce-fir forest, for example) will likely cause a decline in populations in that habitat and therefore an increase in vulnerability of those populations, and may result in their extinctions. A reduction in the number of habitats will likely result in abrupt extinctions.

3. Can individual species be recognized as especially vulnerable a priori?

Again, the answer is clearly yes. Researchers have demonstrated that within the pattern of the diversity-area correlation abides an equally profound pattern (Chapter 10; Brown 1971, 1978; Patterson 1984; Lomolino et al. 1989; Grayson and Livingston 1993). The fauna of the smallest island (or smallest mountaintop) among the islands of a given historically defined region will comprise a given set of species. On the next larger island will be the same species, plus a few more. On the next larger island all of those species will be present, plus a few more, and so on. Bruce Patterson and colleagues have named this the "nested subset" pattern and have investigated it intensively (Chapter 10). The existence of this pattern demonstrates that extinction in these faunas has been a highly regular and nonrandom process: the extinction that has produced the pattern has occurred independently on each island or mountain, yet the pattern is the same on each island.

We can draw two important inferences from the "nested subset" pattern. First, each species requires a minimum area that is necessary to support a viable population. For a species that occurs on a given set of islands, the population that occurs on the smallest island in the set is likely to be most vulnerable to extinction. In the case of the red squirrels in the Southwest, the Pinaleños provide the smallest mountain area on which red squirrels occur (Patterson 1984; Lomolino et al. 1989). Thus, even without data showing the drastic decline in the population in the late 1980s (Chapter 14), biologists could have predicted that they would be vulnerable to habitat destruction and easily pushed to the brink of extinction. The population of the long-tailed vole (*Microtus longicaudus leucophaeus*) in the Pinaleños is in a similarly precarious situation, and, quite probably, so are other animals and some plants. Whether they are legally designated or not, we need no further data to know they are vulnerable or may easily become seriously endangered.

The second inference to be drawn is that other populations on nearby mountains are also vulnerable. We must view the Pinaleños not simply as the best example of a sky island (which it may well be), but as one of a group of unique places. If we are to preserve the best, the most unusual, or the most endangered of the sky-island ecosystems, we must extend protection to these other mountains.

4. Do we have unmistakable evidence that such small populations are actually vulnerable?

One might argue that the diversity-area and nested subset patterns result simply from extrapolation from current distributions and that they present no compelling evidence that extinction has actually taken place under conditions of declining habitat area. However, the fossil record demonstrates that unmistakably real extinctions have taken place (Chapter 10; Grayson and Livingston 1993). Most of the species now confined to the cool, moist mountaintops were much more widely distributed during the cool, moist climate of the last period of glacial advance, which ended about 10,000 years ago. The current distributions of many montane mammals in the Southwest are relicts of better times. In each case, the existence of these fossils confirms the patterns that biologists have deduced on the basis of the living fauna and also confirms both the dynamic nature of distributions and the critical vulnerability of small, isolated populations to environmental perturbations.

5. What places need to be preserved to retain the unique biological features of the sky islands?

To successfully protect the unique biotic communities of the Pinaleños, we must certainly provide protection to the most restricted and fragile of the montane habitats, especially the highest spruce-fir and wet meadow communities. However, the entire elevational gradient must be afforded protection, not just the top. Seasonal movements of many species of birds and mammals are well documented, with even such species as chipmunks often having distinct summer and spring ranges at higher and lower elevations. Moreover, in climatically exceptional years refugia are crucial; such refugia may be in high elevations in dry years or in lower elevations in wet or cold years, for example. Further, interactions among species that have different overall ranges may be crucial; for example, the occasional black bear that forages at the top of the mountain in summer may create critical habitats for plant species when it digs holes in the course of hunting for insects, creating small-scale soil disturbance. All such activities are intrinsic to the functioning of the system and must be retained.

As I noted earlier, we certainly cannot stop at protecting only one of the sky islands. Each will have some unique biotic aspects, and each deserves protection on that basis. Further, it is the diversity of the sky-island ecosystem that is unique and most appealing, and only an array of mountains of varying sizes and heights can preserve that diversity. Such a system of sky-island preserves would not only appeal to the naturalist but would also provide a natural laboratory for deepening and further enhancing our understanding of this already classic region (see Epilogue). Protecting unique places such as the Pinaleños is critical to providing circumstances that permit further development of our understanding of the functioning of habitat-island ecosystems, as well as to preserving its individually unique biota.

6. Which species need attention individually?

There can be no doubt that the Mt. Graham red squirrel is critically vulnerable, due to its very small population size and its irregular usage of portions of the mountain other than high-elevation old-growth forest. Before I discuss this species further, I must emphasize again that other populations are equally and critically vulnerable. Any species on the mountain that occurs exclusively in high-elevation habitats, especially those that tend to occur at low population densities, must be assumed to be vulnerable. But the voles, several invertebrates, and several plants that are unique taxa are also interacting parts of unique communities, and the conspicuous vulnerability of one species is almost certainly paralleled by the vulnerability of other, less conspicuous species. Conservation on the Pinaleños ultimately can be successful only by protecting entire communities, not by focusing exclusively on the individual parts of those communities.

7. What can we do about endangered species?

The best way to protect endangered species is to prevent their becoming endangered. This simple statement is intended as a challenge for conservation activities in this country to shift from *reactive* to *preventive* programs. Responding to crises consumes time and resources that could be used far more effectively in developing policies and procedures that allow human activities to proceed without endangering biological diversity.

This challenge is especially critical to the Forest Service, the Bureau of Land Management, and other federal and state agencies that determine land-use patterns for public lands. By adjusting their general policies, they can have more impact on conservation in the western United States than all other organizations combined. Virtually every organism is affected by two of the most common activities on public lands—logging and grazing, both regulated by these governmental agencies.

In the case of the Mt. Graham red squirrel, it seems certain that logging of old growth has been the primary culprit in the current crisis. (I say this not to point the finger at anyone but to make clear that logging policies are of crucial importance not just in the Pacific Northwest but in all portions of the nation.) Without casting blame, let us now take action that is in keeping with what we know and what we are legally and ethically required to perform. The cessation of logging on Mt. Graham several years ago is probably the only reason that we have any hope of protecting this unique animal.

In a similar fashion, we must recognize that overgrazing on the mountain has severely damaged meadow communities, including the unique voles and plants that constitute them. Overgrazing in all parts of the nation has probably done damage equal to that of overlogging; abundant evidence demonstrates the role of overgrazing in the destruction of freshwater, riparian, and meadow communities. Resource managers have greatly reduced grazing in the Pinaleños and have eliminated it from the highest reaches, but it continues on the other sky islands. Where will we be called to next to manage a crisis that could have been avoided by simple changes in policy?

But when it is too late to avoid the crisis, what can we do? Such actions are far more difficult to describe in general terms than are preventive measures, for each must be tailored to individual species and biotas. Planting trees, bringing in fallen logs to help provide cover and sites for middens, and even such heroic measures as providing artificial sources of food seem worthy of investigation. For the voles, reseeding meadows with native plants and perhaps controlled burns are likely prospects. Whatever is done, I urge that it be done in a carefully documented fashion and made available in published form, so that we may all learn from the process, and not be buried in inaccessible, unpublished reports.

What Next for Conservation Biology in the Pinaleños?

What do I, as an outsider, see as important needs for protection of the Pinaleños communities, especially with regard to current programs? First, long-term monitoring of not just the squirrels but the entire biotic system is badly needed for broadening our understanding of sky-island ecosystems and how to protect them. The Pinaleños have the potential to become the classic case study of biological diversity in the entire Southwest. As part of this effort, biologists should conduct what may be called retrospective monitoring through land-use surveys, dendrochronology, the examination of fossil remains, and other research tools (Chapters 7, 10, 15). The history of land use is crucial to all aspects of interpretation; especially, the history of past grazing, logging, and road building needs full documentation as a context for other studies. Dendrochronology and fossil remains can provide key information on the ecology of the mountain long before Europeans began to alter it (Chapters 7 and 10).

A critical aspect of this work will be documenting the role of environmental variation. We have heard about the possibility of a 20-year drought cycle. This cycle, if it exists, would obviously be crucial both to the survival of red squirrels and to our understanding of these ecosystems and their conservation in the long run. Future extinctions might very well be caused by future broad-scale environmental modifications such as rising global temperatures, acid rain, and air pollution or by local modifications such as fire suppression and creation of edge effects around clearings and roads (Wilcove et al. 1986). Baseline data from the Pinaleños could well provide a critical base of comparison for interpreting and responding to future crises.

However, I wish to emphasize that although the need in the Pinaleños is great (particularly for the endangered species that are present), we must maintain a broad perspective about how the new knowledge may be applied over the entire region, so that the research does not become a sterile exercise, carried out in isolation from management. Biologists must conduct their research properly and publish all relevant information in refereed journals where all concerned parties have free access to the information and may be assured that conclusions are drawn on the basis of rigorously tested data sets.

Success in protecting the Pinaleños and other sky islands will almost certainly depend on the continuation of interaction and cooperation of what are sometimes viewed as discrete camps in academia and management. At the workshop were people from both groups and several who fell in between. There I expressed the hope that a set of specific recommendations would emerge from each of the academic specialists about individual activities on the mountain, followed by presentation of discrete plans by the managers that take advantage of those recommendations; the degree to which this has and has not happened is evident in this volume. Definite action is still needed as soon as possible, and cooperation is still essential.

The Observatory and the Squirrels

All of the circumstances that I have described, all of the patterns of extinction that I have documented, make it abundantly clear that the

Mt. Graham red squirrel was a vulnerable population before humans arrived in Arizona. However, human actions (particularly over the past hundred years) have made them progressively more vulnerable. Today they are unquestionably highly endangered, with their survival dependent both on all of the best assistance that we as conservationists can provide and on simple luck. The removal of habitat at this juncture for any purpose, including construction of a telescope complex, would almost certainly have a negative impact and, I believe, would on biological grounds be irresponsible and indefensible.

The Observatory and the Law

Finally, I wish to return to the topic of interactions between conservation and legal systems. As I noted earlier, the Mt. Graham red squirrel has become a symbol for the high-elevation biota of the mountain because it is the only animal in the Pinaleños currently given protection by the Endangered Species Act (ESA). Because the ESA is virtually the only legal means of rapidly (and I use that term in a relative sense) responding to critical problems, it has become the central legislation in a matrix of complex legal interactions between groups with differing points of view (Chapters 1–3).

We live in a society that depends for its existence on the acceptance of the view that the law prevails and applies equally to all. Our society begins to break down when a portion of the public perceives that their rights are superior or inferior to those of others. For this reason I, as both a citizen and a scientist, abhor the efforts of the University of Arizona to hold itself above the law. When the university petitioned Congress to exempt it from the ESA, it in effect stated that the law should not be adhered to by the university or, by implication, by others. Given the tremendous importance of the ESA and the number of groups that rely on it as the basis for resolving their strongly differing interests within the legal system, the university has opened the floodgates that formerly prevented other powerful and wealthy organizations from disregarding the public (environmental) good. Such an action can only be destructive. At the workshop, I and others called on the representatives of the university to commit themselves to acting in a responsible and ethical manner within the confines of the law. The continued construction of the observatory has revealed the extent to which the university is an ethical organization and is willing to function in a system that is open and fair to all.

References

Brown, J. H. 1971. Mammals on mountaintops: Nonequilibrium insular biogeography. *American Naturalist* 105:467–78.

Brown, J. H. 1978. The theory of insular biogeography and the distribution of boreal birds and mammals. *Great Basin Naturalist Memoirs* 2:209–27.

Fairbanks, R. G. 1989. A 17,000-year glacio-eustatic sea level record: Influence of glacial melting rates on the Younger Dryas event and deep-ocean circulation. *Nature* 342:637–42.

Grayson, D. K., and S. D. Livingston. 1993. Missing mammals on Great Basin Mountains: Holocene extinctions and inadequate knowledge. *Conservation Biology* 7:527–32.

Heaney, L. R. 1984a. Mammalian species richness on islands on the Sunda Shelf, Southeast Asia. *Oecologia* 61:11–17.

Heaney, L. R. 1984b. Climatic influences on life history tactics and behavior of North American tree squirrels. In *The Biology of Ground Dwelling Squirrels,* ed. J. O. Murie and G. R. Michener, pp. 43–78. University of Nebraska Press, Lincoln.

Heaney, L. R. 1986. Biogeography of mammals in Southeast Asia: Estimates of rates of colonization, extinction, and speciation. *Biological Journal of the Linnean Society* 28:127–65.

Heaney, L. R. 1991a. A synopsis of climatic and vegetational change in Southeast Asia. *Climatic Change* 19:53–61.

Heaney, L. R. 1991b. An analysis of patterns of distribution and species richness among Philippine fruit bats (Pteropodidae). *Bulletin of the American Museum of Natural History* 206:145–67.

Heaney, L. R. 1993. Biodiversity patterns and the conservation of mammals in the Philippines. *Asia Life Sciences* 2:261–74.

Heaney, L. R., and E. A. Rickart. 1990. Correlations of clades and clines: Geographic, elevational, and phylogenetic distribution patterns among Philippine mammals. In *Vertebrate Biogeography and Speciation in the Tropics,* ed. G. Peters and R. Hutterer. Museum of Alexander Koenig, Bonn.

Lomolino, M. V., J. H. Brown, and R. Davis. 1989. Island biogeography of montane forest mammals in the American Southwest. *Ecology* 70:180–94.

Musser, G. G. 1988. The mammals of Sulawesi. In *Biogeographical Evolution of the Malay Archipelago,* ed. T. C. Whitmore, pp. 73–93. Clarendon Press, Oxford.

Patterson, B. D. 1984. Mammalian extinction and biogeography in the southern Rocky Mountains. In *Extinctions,* ed. M. H. Nitecki, pp. 247–94. University of Chicago Press, Chicago.

Patterson, B. D., and W. Atmar. 1986. Nested subsets and the structure of insular mammalian faunas and archipelagos. *Biological Journal of the Linnean Society* 28:65–82.

Wilcove, D. S., C. H. McLellan, and A. P. Dobson. 1986. Habitat fragmentation in the temperate zone. In *Conservation Biology: The Science of Scarcity and Diversity,* ed. M. E. Soule, pp. 237–56. Sinauer Press, Sunderland, Mass.

CHAPTER TWELVE

Population Genetics and Conservation Biology of Relict Populations of Red Squirrels

Robert Miles Sullivan and Terry L. Yates

Tree squirrels in the genus *Tamiasciurus* have until recently included only two species: the Douglas' squirrel (*T. douglasii*) and the red squirrel (*T. hudsonicus*) (Hall 1981). Populations of red squirrels are distributed throughout most coniferous forest habitats in North America. Although the geographic distribution of each species is essentially continuous, several populations of *T. hudsonicus* inhabit isolated mountain ranges in Arizona and New Mexico, and a relict population of *T. douglasii* occurs in the Sierra San Pedro Martir of Baja California. Using statistical analysis of cranial morphology, S. L. Lindsay (1982) elevated the Baja California population of *T. douglasii* to full species status (*T. mearnsi*). This morphological analysis separated *T. hudsonicus* and *T. douglasii* from the disjunct Baja California group by equal morphologic intervals and taxonomic distance. However, because this analysis was based entirely on morphologic gaps of equal size among groups, the systematic and evolutionary relationships among the three taxa remain problematical.

In addition, biologists recently have focused considerable attention on the taxonomic status of *T. h. grahamensis* in the Pinaleño Mountains of southern Arizona. This endangered subspecies of red squirrel represents the southernmost marginal subspecies of *T. hudsonicus* in western North America (figure 12.1; Hall 1981). It is restricted to a small area of forest atop Mt. Graham (Chapter 14).

Herein, we describe genetic variation and reassess the systematic relationships among red squirrels in the genus *Tamiasciurus*. Genetic data are used as an independent test of the morphologic species concept (Sokal and Crovello 1970; Lindsay 1982) and to describe genetic structure and patterns of microevolutionary differentiation among isolated

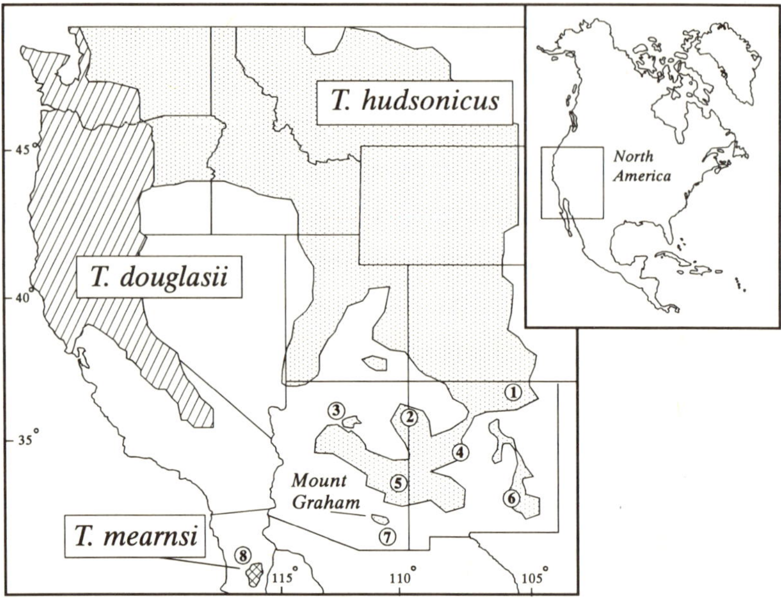

Figure 12.1 Map of study area showing the geographic distribution of sample localities. Diagonal lines = *Tamiasciurus douglasii;* dense stipple = *T. hudsonicus;* crosshatched = *T. mearnsi.* 1 = southern Rocky Mountains, 2 = Chuska Mountains, 3 = Kaibab Plateau, 4 = Mt. Taylor–Zuni Mountain complex, 5 = White-Mogollon Mountain complex, 6 = Capitan-Sacramento Mountain complex, 7 = Mt. Graham, and 8 = Sierra San Pedro Martir.

populations of *T. hudsonicus* in the American Southwest. Our analysis focuses specifically on the disjunct population of *T. douglasii* (cited in the analysis as *T. mearnsi*) in Baja California and the relict population of *T. h. grahamensis* in the Pinaleño Mountains of southeastern Arizona. By comparing the relative degree of genetic divergence exhibited by populations of *Tamiasciurus* at different hierarchical (taxonomic) levels, we should be better able to discern the uniqueness of populations in relation to current systematic and conservation issues. If these populations are, in fact, genetically unique, their gene pools may be worthy of special protection.

Methods

Protein Electrophoresis

We assayed genetic variability for 26 presumptive gene loci using starch-gel electrophoresis (Selander et al. 1971; Harris and Hopkinson

1976) and designated genetic loci, isozyme products of more than one locus, and electromorphs (following Richardson et al. 1986). We then analyzed genotypes using BIOSYS-l (Swofford and Selander 1981) to determine (1) allelic frequencies, (2) measures of genetic variability (i.e., the average number of alleles per locus [*A*], percentage of the loci that were polymorphic [*P*], and mean heterozygosity [*H*-obs.], expected heterozygosity [*H*-exp.]), and (3) pairwise genetic similarities (S-values; Rogers 1972) among populations of each species.

Phenetic Analysis

Because coniferous forest populations of *Tamiasciurus* in the Southwest have relict, islandlike distributions, we assumed that divergence among allopatric (isolated) populations occurred primarily from the process of random genetic drift. Thus, we calculated genetic distances (following Cavalli-Sforza and Edwards 1967) from allelic data for each pair of populations (Wright 1978; Nei et al. 1983; Rogers 1984; Wayne and O'Brien 1987; Swofford and Olsen 1990). This measure of minimum evolution avoids objections (Farris 1981) concerning the use of genetic distance measures in constructing phylogenetic trees (Swofford 1981; Rogers 1984). We then assessed phylogenetic relationships among species and populations using the UPGMA cluster analysis (Swofford and Selander 1981; Wayne and O'Brien 1987; Honeycutt and Wilkinson 1989) on the genetic distance matrix (Swofford 1981; Rogers 1984). Voucher specimens were deposited in the Museum of Southwestern Biology at the University of New Mexico.

Results

Intrapopulation Variability

Allelic frequencies, average number of alleles per locus (*A*), average polymorphism (*P*), observed heterozygosity (*H*-obs.), and expected heterozygosity per population (*H*-exp.) for the six polymorphic loci are given in tables 12.1 and 12.2. Of the 26 loci examined, 18 were monomorphic (not variable) in all groups. Estimates of intrapopulation variability among species of *Tamiasciurus* were much lower than values reported for species of other small mammals (table 12.2; Nevo 1978; Sullivan 1985; Sullivan and Petersen 1988; Sullivan 1994). Among the three species of *Tamiasciurus* the average number of alleles per locus ranged from 1.00 to 1.10, the average percentage of loci polymorphic ranged from 0.00 to 7.70, and the average heterozygosity observed

Table 12.1 Allele Frequencies of Polymorphic Loci

Locus/Allele		*T. douglasii* (18)	*T. mearnsi* (5)	Populations of *Tamiasciurus hudsonicus*						
				White-Mogollon Mts. (17)	Kaibab Plateau (9)	Chuska Mts. (14)	Southern Rocky Mts. (24)	Mt. Taylor-Zuni Mts. (7)	Capitan-Sacramento Mts. (19)	Mt. Graham (5)
Alb	a	1.000	1.000	0.000	0.000	0.000	0.000	0.000	0.000	0.000
	b	0.000	0.000	1.000	1.000	1.000	1.000	1.000	1.000	1.000
Me-2	a	0.000	0.000	0.059	0.000	0.000	0.000	0.000	0.000	0.000
	b	1.000	1.000	0.941	1.000	1.000	1.000	1.000	1.000	1.000
Pgd	a	0.028	0.000	0.000	0.000	0.000	0.000	0.000	0.000	0.000
	b	0.972	1.000	1.000	1.000	1.000	1.000	1.000	1.000	1.000
Got-1	a	0.056	0.000	0.000	0.000	0.000	0.000	0.000	0.000	0.000
	b	0.944	1.000	1.000	1.000	1.000	1.000	1.000	1.000	1.000
Pgm-2	a	1.000	1.000	0.000	0.000	0.000	0.000	0.000	0.000	0.000
	b	0.000	0.000	1.000	1.000	1.000	1.000	1.000	1.000	1.000
Pep-C	a	1.000	1.000	0.882	1.000	0.929	1.000	1.000	1.000	1.000
	b	0.000	0.000	0.118	0.000	0.071	0.000	0.000	0.000	0.000
Pep-B	a	1.000	1.000	1.000	0.833	0.750	1.000	1.000	1.000	1.000
	b	0.000	0.000	0.000	0.167	0.250	0.000	0.000	0.000	0.000
Ada	a	0.000	0.000	0.029	0.000	0.000	0.000	0.000	0.000	1.000
	b	1.000	1.000	0.971	1.000	1.000	1.000	1.000	1.000	0.000

Note: Sample sizes are indicated in parentheses.

Table 12.2 Genetic Variability at 26 Loci in *Tamiasciurus douglasii*, *T. mearnsi*, and allopatric populations of *T. hudsonicus*

Species/Population	*N*	*A*	*P*	*H*-obs	*H*-exp*
Tamiasciurus douglasii	18	1.1	7.7	0.002 (0.002)	0.006 (0.005)
Tamiasciurus mearnsi	5	1.0	0.0	0.000 (0.000)	0.000 (0.000)
Tamiasciurus hudsonicus					
White-Mogollon Mts.	17	1.1	11.5	0.007 (0.005)	0.015 (0.009)
Kaibab Plateau	9	1.0	3.8	0.013 (0.013)	0.011 (0.011)
Chuska Mts.	14	1.1	7.7	0.019 (0.019)	0.020 (0.016)
Southern Rocky Mts.	24	1.0	0.0	0.000 (0.000)	0.000 (0.000)
Mt. Taylor–Zuni Mts.	7	1.0	0.0	0.000 (0.000)	0.000 (0.000)
Capitan-Sacramento Mts.	19	1.0	0.0	0.000 (0.000)	0.000 (0.000)
Mt. Graham	5	1.0	0.0	0.000 (0.000)	0.000 (0.000)

Note: Mean sample size = *N*; mean number alleles per locus = *A*; percentage of loci polymorphic = *P*; average heterozygosity observed = *H*-obs and expected = *H*-exp.; standard errors are in parentheses. A locus was considered polymorphic if more than one allele was detected in the sample. * = unbiased estimate (Nei 1978).

ranged from 0.000 to 0.041. Levels of intrapopulation genetic variation in *T. hudsonicus* also were extremely low. The White-Mogollon Mountain complex followed by the Chuska Mountains and Kaibab Plateau samples exhibited only minor amounts of genetic variation, whereas samples of *T. hudsonicus* from the southern Rocky Mountains, Mt. Taylor–Zuni Mountain complex, Capitan-Sacramento Mountain complex, and Pinaleño Mountains exhibited no intraspecific genetic variation at the 26 enzyme loci examined (table 12.1).

Average genetic similarity among all three species of red squirrels was high (S values of 0.881–1.000; table 12.3). However, *T. mearnsi* was more similar to *T. douglasii* (S value of 0.997) than either of those species was to *T. hudsonicus* (S values of 0.908–0.923). Average genetic similarity among allopatric populations of *T. hudsonicus* also was high (S values of 0.949–1.000, including *T. h. grahamensis*). Among disjunct populations of *T. hudsonicus*, the most disparate similarity

Table 12.3 Matrix of Genetic Similarity Values and Genetic Distances

Species/Population	*T. douglasii* (18)	*T. mearnsi* (5)	White-Mogollon Mts. (17)	Kaibab Plateau (9)	Chuska Mts. (14)	Southern Rocky Mts. (24)	Mt. Taylor-Zuni Mts. (7)	Capitan-Sacramento Mts. (19)	Mt. Graham (5)
Tamiasciurus douglasii	—	0.997	0.912	0.913	0.908	0.920	0.920	0.920	0.881
Tamiasciurus mearnsi	0.036	—	0.915	0.917	0.911	0.923	0.923	0.923	0.885
White-Mogollon Mts.	0.259	0.256	—	0.986	0.985	0.992	0.992	0.992	0.956
Kaibab Plateau	0.258	0.255	0.077	—	0.994	0.994	0.994	0.994	0.955
Chuska Mts.	0.263	0.260	0.075	0.036	—	0.988	0.988	0.988	0.949
Rocky Mts.	0.256	0.250	0.057	0.052	0.073	—	1.000	1.000	0.962
Mt. Taylor-Zuni Mts.	0.256	0.250	0.057	0.052	0.073	0.000	—	1.000	0.962
Capitan-Sacramento Mts.	0.256	0.250	0.057	0.052	0.073	0.000	0.000	—	0.962
Mt. Graham	0.308	0.306	0.169	0.184	0.191	0.177	0.177	0.177	—

Note: Genetic similarity values (S-values, above diagonal) are based on Rogers 1972; genetic distance (CORD, below diagonal) are based on Cavalli-Sforza and Edwards 1967. Sample sizes are indicated in parentheses.

values were between the Pinaleños sample (*T. h. grahamensis*) and all other sample localities (S values of 0.949–0.962).

Distance Analysis and Phyletic Relationships

In the eight polymorphic loci assayed, we identified 10 allelic variants among the three species of *Tamiasciurus,* but only four loci were polymorphic among allopatric populations of *T. hudsonicus* (table 12.1). *Tamiasciurus douglasii* and *T. mearnsi* were fixed for (i.e., present in all individuals) two unique alleles (Alb^a, $Pgm\text{-}2^a$), whereas all populations of *T. hudsonicus* were fixed for different unique alleles at the same two loci (Alb^b, $Pgm\text{-}2^b$) (figure 12.2). Except for two rare alleles (Pgd^a and $Got\text{-}1^a$), *T. douglasii* and *T. mearnsi* were genetically identical. These allelic data and the UPGMA phenogram (figure 12.2) provide no evidence to corroborate the morphometric hypothesis that the disjunct Baja California population of *Tamiasciurus* warrants recognition as a distinct species (Lindsay 1982). Instead, we recommend that the Sierra San Pedro Martir population retain its subspecific status (*T. d. mearnsi*), which was based originally on morphological criteria (Allen 1894; Lindsay 1982; Hoffmeister 1986).

Among allopatric populations of *T. hudsonicus,* one allele was shared between the White-Mogollon Mountain complex and the Chuska Mountains ($Pep\text{-}C^b$); one allele was shared between the Chuska Mountains and the Kaibab Plateau sample ($Pep\text{-}B^b$); one allele (Ada^a) was shared between the Pinaleño (Mt. Graham) sample and the White-Mogollon Mountain complex; and one allele was unique but not fixed in all individuals from the White-Mogollon Mountain complex ($Me\text{-}2^a$). The UPGMA phenogram (figure 12.2) reveals three salient features. First, phyletic affinities among populations of red squirrels from the southern Rocky Mountains, Mt. Taylor–Zuni Mountain complex, and Capitan-Sacramento Mountain complex are unresolved. The White-Mogollon Mountain sample is a close outlier to this homogeneous group but differs somewhat because of the presence of one unique rare allele ($Me\text{-}2^a$) and because it shares an allele ($Pep\text{-}C^b$) in low frequency with the Chuska Mountains (table 12.1; figure 12.2). Second, a sister-site relationship is indicated between the Kaibab Plateau and the Chuska Mountain samples because these two sites share an allele ($Pep\text{-}B^b$) in low frequency. Third, because the Pinaleños sample was fixed for Ada^a in the five specimens assayed, it is represented in the UPGMA phenogram as the most highly differentiated population of

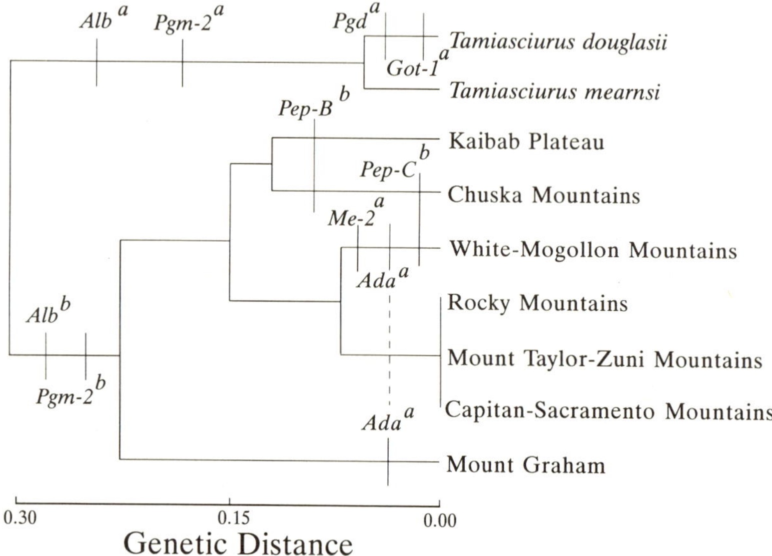

Figure 12.2 UPGMA phenogram of cord distances (Cavalli-Sforza and Edwards 1967) showing allozyme relationships among groups of *Tamiasciurus*. The cophenetic correlation coefficient = 0.993, and percent standard deviation = 9.631. Polymorphic alleles are superimposed on the phenogram.

red squirrel analyzed. This allele, however, was found in the heterozygous state in one specimen from the White-Mogollon Mountain complex, suggesting that gene flow may have occurred between these two biogeographic sites during the Pleistocene.

Discussion

Genetic Variation and Biogeographic Patterns

Tamiasciurus hudsonicus grahamensis was originally described (Allen 1894) as a distinct subspecies based on coloration, size, and degree of geographic isolation (Allen 1894, 1895, 1898). Despite several taxonomic and nomenclatural revisions, this subspecies has retained its taxonomic status in the two most recent faunal accounts (Hall 1981; Hoffmeister 1986); our analyses of allelic variation in *T. hudsonicus* also support the subspecific status of the Pinaleños population.

At least two explanations are possible for the relative lack of allelic divergence in populations of red squirrels: dispersal along mesic woodland corridors and failure to differentiate genetically. First, populations of red squirrels may be insensitive to existing geographic discontinuities because they could disperse along mesic woodland corridors that surround and connect contemporary coniferous forest populations in this area—these sites are connected by corridors of Great Basin Coniferous Forest or Woodland macrohabitat until about 2,300 m in elevation (Brown and Lowe 1980; Brown 1982). Moreover, inspection of the two most recent reconstructions of the hypothesized distribution of late Pleistocene macrohabitat in the Southwest (Lomolino et al. 1989: 186; Harris 1990:221; A. H. Harris personal communication 1994) indicates that mountaintops in eastern Arizona and north-central New Mexico (i.e., the White-Mogollon Mountain complex, Mt. Taylor–Zuni Mountain complex, San Mateo Mountains, and Black Range but with the possible exceptions of the Chuska Mountains and the Magdalena Mountains in Arizona and New Mexico, respectively) were surrounded and connected by spruce-fir forests until the late Pleistocene (15,000–11,000 years ago)—thus allowing for dispersal of populations and gene flow. Conversely, all other mountaintops in southern Arizona and southwestern New Mexico were separated at low elevations within or below mixed-conifer forest habitat and, therefore, do not constitute a historically homogeneous subset of sites conducive to dispersal by red squirrels and other montane species.

Thus, patterns of genetic differentiation in *T. hudsonicus* residing in eastern Arizona and north-central New Mexico may be the result of post-Pleistocene dispersal and subsequent gene flow among relict populations located in more northern latitudes. In contrast, desert grassland macrohabitat, in concert with founder effects (i.e., a small number of dispersing individuals representing a biased genetic sample of the parental population), may have acted as an effective filter barrier to gene flow among populations inhabiting more southern latitudes, including the Pinaleños (Davis et al. 1988; Lomolino et al. 1989; Sullivan 1994). Paleobotanical and fossil vertebrate data indicate that between 10,000 and 30,000 years ago many forests were isolated by woodlands, especially in southeastern Arizona and southwestern New Mexico (Lomolino et al. 1989; Harris 1990). Some of these more southern mountaintops, therefore, must have been colonized by dispersing coniferous forest species during late Pleistocene and early Holocene times (Davis and Dunford 1987; Davis et al. 1988).

The Pinaleño Mountains, one of these more southerly mountain ranges, are separated from the nearest highlands to the north by the Gila River Valley. Dispersal from the White-Mogollon Mountain complex or the Black Range south to the Animas-Peloncillo highland and then northwest to the Pinaleño Mountains would not have required crossing a gap between mixed-conifer forest (Lomolino et al. 1989) or middle-elevation savanna vegetation greater than about 21 km (Harris 1990). If, however, dispersing coniferous forest species followed the continental divide, the much broader low-elevation gap may have acted as a filter barrier (i.e., allowing access to some species, while denying access to others) to highland forms such as *Tamiasciurus* (Harris 1990). If the contemporary distribution of *Tamiasciurus* is correct, this species must have gone extinct in intermediate areas of coniferous forest habitat (e.g., the Chiricahua and Animas Mountains); thus, occasional colonization followed by local Holocene extinctions appears to be the most likely biogeographical scenario (Harris 1990). This dispersal hypothesis is supported by the presence of an allele (Ada^a) found in the heterozygous state in only one red squirrel from the White-Mogollon Mountain complex; it was fixed, however, in the five available specimens assayed from the Pinaleños (figure 12.2). Although inferential, these data suggest that populations of red squirrels from the Mogollon Rim may have been the ancestral source from which dispersing individuals ultimately colonized the Pinaleño Mountains.

Second, the resolving power of protein electrophoresis indicates considerable genetic similarity among geographically distant populations of red squirrels; this species simply may have failed to differentiate in response to late Holocene vicariant events (<4,000 years ago). Consequently, researchers may be able to better assess the divergence and timing of dispersal events in the genus *Tamiasciurus* using restriction fragment analysis of mitochondrial DNA (mt-DNA), or DNA amplification (PCR) techniques (Pumo et al. 1988; Riddle and Honeycutt 1990).

Genetics and Conservation of Relict Montane Populations

Small relict populations are threatened by at least three types of chance phenomena, including genetic, demographic, and environmental stochasticity (Shatter 1981; Simberloff 1988). Increasing evidence suggests that genetic diversity is positively associated with increased disease resistance, growth rate, and developmental stability in animals (Allendorf and Leary 1988; Murphy et al. 1990). Conservation scientists in general agree that genetic diversity is a nonrenewable resource that must be preserved to increase the probability of both short- and long-term survival of animal species (Allendorf 1986; Meffe 1986, 1987). Maintaining existing levels of genetic diversity within populations of threatened or endangered taxa is an important component of resource management, which cannot easily be separated from the resource of biotic diversity (Allendorf and Phelps 1981; Turner 1983; Vrijenhoek et al. 1985; Meffe 1986, 1987; Echelle et al. 1989).

A growing concern in the management of endangered wildlife is the increasing insularity and habitat fragmentation of disjunct populations. Inbreeding, in conjunction with population subdivision through habitat fragmentation, may affect the level of genetic diversity in small populations and may alter the evolutionary potential for optimal partitioning of energy into processes associated with development and reproduction (Naso et al. 1974; Lynch 1977; Cothran et al. 1983). Because *T. hudsonicus* in the Southwest occurs in a variety of widely separated areas, the species as a whole would seem buffered against catastrophic extinction. However, all populations peripheral to the southern Rocky Mountains inhabit relatively small areas of coniferous forest habitat; about three-quarters of the genetic variation found in red squirrels is attributable to differentiation in these small isolated populations. The genetically depauperate state of these populations, in conjunction with their demographic and ecological characteristics, may make them particularly susceptible to catastrophic extinction owing to the simultaneous effects of human-induced habitat destruction, human disturbance, disease (e.g., hantavirus infection), and the vagaries of environmental stochasticity. Conservation biologists can use electrophoretic assays as well as other genetic data to identify unique gene pools worthy of protection or special management consideration, particularly when unique alleles are present in one population but absent in others.

In the Southwest, a goal of resource management must be to preserve biological diversity in these relict coniferous forest ecosystems. This remaining diversity can provide irreplaceable biogeographic links to the unique evolutionary history of the Southwest. Given the accelerating habitat destruction and fragmentation of forest habitats, resource managers must make a special effort to preserve as much of this delicate species pool as possible. Like many mountain ranges in the Southwest, the Pinaleño Mountains are sanctuary to a complex mosaic of mountaintop environments, harboring a diversity of unique plant and animal life. As intra-island habitat patches become more isolated, fragmented, or shredded (Feinsinger 1994) through intervening habitat destruction, the search for seasonal foods beyond a certain distance threshold makes endemic species vulnerable to starvation, predation, and disease (Soule 1986). Species with patchy distributions or taxa that use a variety of microhabitats are particularly vulnerable to extinction in a fragmented landscape (Wilcove et al. 1986). Indeed, for species restricted to islandlike habitats with no opportunity for recolonization, population viability analysis should anticipate the worst-case scenario for long-term survival (Gilpin and Soule 1986).

Our analysis suggests that the Mt. Graham red squirrel is a distinct lineage that warrants protection. Our data are also consistent with the previously proposed hypothesis that the Pinaleño Mountains may have been one of the first of many mountaintops in the Southwest to become isolated (Sullivan 1994). As such, red squirrels on Mt. Graham may represent the "tip of the iceberg" in terms of biological diversity unique to this locality. Considered in this context, preservation of this lineage of red squirrel becomes even more important. Moreover, preservation of the historical legacy and natural pattern of genetic diversity through habitat management in one species can contribute significantly toward conservation of other members of the biota, which often are local endemics that have shared a common history (Echelle et al. 1989). The Pinaleño Mountains may contain many endemic, protected, or rare species of both plants and animals. Researchers have yet to study many aspects of the Pinaleños ecosystem in a contemporary or historical context. The Pinaleño Mountains represent a unique Southwest coniferous forest environment that deserves protection.

Acknowledgments

We thank T. L. Best, T. E. Lawlor, L. J. Janecek, B. D. Patterson, and particularly K. E. Petersen and S. D. Sullivan for their constructive comments or invaluable assistance in the field. Financial support for this study was provided by the Endangered Species Program of the New Mexico Department of Game and Fish, the National Science Foundation (grant Nos. DEB-8004685 and BSR 8501209), the National Institutes of Health (DRR-RR08139-14), the American Society of Mammalogy, and the Buffalo Foundation.

References

Allen, J. A. 1894. Descriptions of ten new North American mammals, and remarks on others. *Bulletin of the American Museum of Natural History* 6:320–21.

Allen, J. A. 1895. On a collection of mammals from Arizona and Mexico made by Mr. W. W. Price, with field notes by the collector. *Bulletin of the American Museum of Natural History* 7:193–258.

Allen, J. A. 1898. Revision of the chickarees, or North American red squirrels (subgenus *Tamiasciurus*). *Bulletin of the American Museum of Natural History* 10:249–98.

Allendorf, F. W. 1986. Genetic drift and the loss of alleles versus heterozygosity. *Zoo Biology* 5:181–90.

Allendorf, F. W., and R. F. Leary. 1988. Conservation and distribution of genetic variation in a polytypic species, the cutthroat trout. *Conservation Biology* 2:170–84.

Allendorf, F. W., and S. R. Phelps. 1981. Isozyme and the preservation of genetic variation in salmonid fishes. In *Fish Gene Pools,* ed. N. Ryman, pp. 37–52. Ecological Bulletins, Stockholm.

Brown, D. E. 1982. Biotic communities in the American Southwest–United States and Mexico. *Desert Plants* 4:1–342.

Brown, D. E., and C. H. Lowe. 1980. *Biotic Communities of the Southwest.* General Technical Report RM-78. U.S. Department of Agriculture, Forest Service, Rocky Mt. Forest and Range Experimental Station, Tempe, Ariz.

Cavalli-Sforza, L. L., and W. A. F. Edwards. 1967. Phylogenetic analysis and estimation procedures. *Evolution* 21:550–70.

Cothran, E. G., R. C. Chesser, M. H. Smith, and P. E. Johns. 1983. Influences of genetic variability and maternal factors on fetal growth in white-tailed deer. *Evolution* 37:282–91.

Davis, R., and C. Dunford. 1987. An example of contemporary colonization by small non-flying mammals of montane islands in the American Southwest. *American Naturalist* 129:398–406.

Davis, R., C. Dunford, and M. V. Lomolino. 1988. Montane mammals of the American Southwest: The possible influences of post-Pleistocene colonization. *Journal of Biogeography* 15:841–48.

Echelle, A. F., A. A. Echelle, and D. R. Edds. 1989. Conservation genetics of a spring-dwelling desert fish, the Pecos Gambusia (*Gambusia nobilis,* Poeciliidae). *Conservation Biology* 3:159–69.

Farris, J. S. 1981. Distance data in phylogenetic analysis. In *Advances in Cladistics: Proceedings of the First Meeting of the Willi Hennig Society,* ed. V. A. Funk and D. R. Brooks, pp. 3–23. New York Botanical Garden, the Bronx.

Feinsinger, P. 1994. Habitat "shredding." In *Principles of Conservation Biology,* ed. G. K. Meffe and C. R. Carrol, pp. 258–60. Sinauer Associates, Sunderland, Mass.

Gilpin, M. E., and M. E. Soule. 1986. Minimum viable populations: Processes of species extinction. In *Conservation Biology: The Science of Scarcity and Diversity,* ed. M. E. Soule, pp. 19–34. Sinauer Associates, Sunderland, Mass.

Hall, E. R. 1981. *The Mammals of North America.* 2d ed. John Wiley and Sons, New York.

Harris, A. H. 1990. Fossil evidence bearing on southwestern mammalian biogeography. *Journal of Mammalogy* 71:219–29.

Harris, H., and A. D. Hopkinson. 1976. *Handbook of Enzyme Electrophoresis in Human Genetics.* American Elsevier Publishing Co., New York.

Hoffmeister, D. F. 1986. *Mammals of Arizona.* University of Arizona Press, Tucson, and the Arizona Game and Fish Department, Phoenix.

Honeycutt, R. L., and P. Wilkinson. 1989. Electrophoretic variation in the parthenogenetic grasshopper *Warramaba virgo* and its sexual relatives. *Evolution* 43:1027–44.

Lindsay, S. L. 1982. Taxonomic and biogeographic relationships of Baja California chickarees (*Tamiasciurus*). *Journal of Mammalogy* 62:673–82.

Lomolino, M. V., J. H. Brown, and R. Davis. 1989. Island biogeography of montane forest mammals in the American Southwest. *Ecology* 70:180–94.

Lynch, C. B. 1977. Inbreeding effects upon animals derived from a wild population of *Mus musculus. Evolution* 31:526–37.

Meffe, G. K. 1986. Conservation genetics and the management of endangered fishes. *Fisheries* 11:14–23.

Meffe, G. K. 1987. Conserving fish genomes: Philosophies and practices. *Environmental Biology of Fishes* 18:3–9.

Murphy, R. W., J. W. Sites Jr., D. G. Buth, and C. H. Haufler. 1990. Proteins I: Isozyme electrophoresis. In *Molecular Systematics,* ed. D. M. Hillis and Craig Moritz, pp. 45–126. Sinauer Associates, Sunderland, Mass.

Naso, R. B., et al. 1975. Weight gain and heterosis at different stages of development of the rat. *Growth* 39:345–61.

Nei, M. 1978. Estimates of average heterozygosity and genetic distance from a small number of individuals. *Genetics* 89:583–90.

Nei, M., F. Tajima, and A. Tateno. 1983. Accuracy of estimated phylogenetic trees from molecular data. *Journal of Molecular Evolution* 19:153–70.

Nevo, E. 1978. Genetic variation in natural populations: Patterns and theory. *Theoretical Population Biology* 13:121–77.

Pumo, D. E., E. Z. Goldin, B. Elliot, C. J. Phillips, and H. H. Genoways. 1988. Mitochondrial DNA polymorphism in three Antillean island populations of the fruit bat, *Artibeus jamaicensis. Molecular Biology and Evolution* 5:79–89.

Richardson, B. J., P. R. Baverstock, and M. Adams. 1986. *Allozyme Electrophoresis: A Handbook for Animal Systematics and Population Studies.* Academic Press, San Diego, Calif.

Riddle, B. R., and R. L. Honeycutt. 1990. Historical biogeography in North American arid regions: An approach using mitochondrial-DNA phylogeny in grasshopper mice (genus *Onychomys*). *Evolution* 44:1–15.

Rogers, J. S. 1972. Measures of genetic similarity and genetic distance. Studies in Genetics 7. *University of Texas Publication* 7213:145–53.

Rogers, J. S. 1984. Deriving phylogenetic trees from allele frequencies. *Systematic Zoology* 33:52–63.

Selander, R. K., M. H. Smith, S. Y. Yang, W. E. Johnson, and J. R. Gentry. 1971. Biochemical polymorphism in the genus *Peromyscus,* part 1, Variation in the old field mouse (*Peromyscus polionotus*). Studies in Genetics 6. *University of Texas Publication* 7103:49–90.

Shaffer, M. L. 1981. Minimum population sizes for species conservation. *Bioscience* 31:131–34.

Simberloff, D. 1988. The contribution of population and community biology to conservation science. *Annual Review of Ecology and Systematics* 19: 473–511.

Sokal, R. R., and T. J. Crovello. 1970. The biological species concept: A critical evaluation. *American Naturalist* 104:127–53.

Soule, M. E. 1986. *Conservation Biology: The Science of Scarcity and Diversity.* Sinauer Associates, Sunderland, Mass.

Sullivan, R. M. 1985. Phyletic, biogeographic, and ecologic relationships among montane populations of least chipmunks (*Eutamias minimus*) in the Southwest. *Systematic Zoology* 34:419–48.

Sullivan, R. M. 1994. Microevolutionary differentiation and biogeographic structure among coniferous forest populations of the Mexican woodrat (*Neotoma mexicana*) in the American Southwest: A test of the vicariance hypothesis. *Journal of Biogeography* 21:369–89.

Sullivan, R. M., and K. E. Petersen. 1988. Systematics of southwestern populations of least chipmunks (*Tamias minimus*) reexamined: A synthetic approach. *Occasional Papers of the Museum of Southwestern Biology, University of New Mexico* 5:1–27.

Swofford, D. L. 1981. On the utility of the distance Wagner procedure. In *Advances in Cladistics: Proceedings of the First Meeting of the Willi Hennig Society,* ed. V. A. Funk and D. R. Brooks, pp. 25–43. New York Botanical Garden, the Bronx.

Swofford, D. L., and G. J. Olsen. 1990. Phylogeny reconstruction. In *Molecular Systematics,* ed. D. M. Hillis and C. Moritz, pp. 411–501. Sinauer Associates, Sunderland, Mass.

Swofford, D. L., and R. B. Selander. 1981. *BIOSYS-l: A Computer Program for the Analysis of Allelic Variation in Genetics.* University of Illinois, Urbana.

Turner, B. J. 1983. Genetic variation and differentiation of remnant populations of the desert pupfish, *Cyprinodon macularius. Evolution* 37:690–700.

Vrijenhoek, R. C., M. E. Douglas, and G. K. Meffe. 1985. Conservation genetics of endangered fish populations in Arizona. *Science* 229:400–402.

Wayne, R. K., and S. J. O'Brien. 1987. Allozyme divergence within the Canidae. *Systematic Zoology* 36:339–55.

Wilcove, D. S., C. H. McLellan, and A. P. Dobson. 1986. Habitat fragmentation in the temperate zone. In *Conservation Biology: The Science of Scarcity and Diversity,* ed. M. E. Soule, pp. 237–56. Sinauer Associates, Sunderland, Mass.

Wright, S. 1978. *Evolution and the Genetics of Populations,* vol. 4, *Variability Within and Among Natural Populations.* University of Chicago Press.

CHAPTER THIRTEEN

The Niche of Diurnal Tree Squirrels

Christopher Smith

The diurnal tree squirrels of the boreal and temperate forests of the Northern Hemisphere are limited to two genera, *Sciurus* and *Tamiasciurus*. One species of *Sciurus, S. vulgaris,* occupies both the boreal and temperate forests of Eurasia, while in North America the genus *Tamiasciurus* occupies boreal forests and the genus *Sciurus* occupies temperate deciduous and southern pine forests (Gurnell 1987). To understand the biology of the Mt. Graham red squirrel (*T. hudsonicus grahamensis*), I will use a comparative approach to demonstrate how differences in the biology of different groups of squirrels adapt the groups to their unique environments. First, I will compare diurnal tree squirrels with ground squirrels and flying squirrels. I will then contrast the biologies of the genera *Sciurus* and *Tamiasciurus* in North America and also contrast the biology of the genus *Tamiasciurus* in boreal forests of North America with that of *S. vulgaris* in boreal forests of Eurasia. Finally, I will contrast the biology of the Mt. Graham red squirrel (*T. hudsonicus grahamensis*) with that of other species and subspecies of *Tamiasciurus*.

The Biology of Diurnal Tree Squirrels and Ground Squirrels

Diurnal tree squirrels feed mainly on the reproductive products of trees and fungi that are relatively rich in lipids (Smith 1968; Smith and Follmer 1972), and they show a preference for foods that are high in lipids (Smith and Follmer 1972). Because plant oils contain many more calories per gram of dry weight (10 to 11 kcal) than do carbohydrates (4 kcal), tree squirrels can gain their daily energy requirements while carrying less weight in their digestive tract by eating foods richer in

Table 13.1 Digestive Efficiency of Three Species of Squirrels

Species of Food	Plant Part Ingested	Ingested Energy (kjoules/g dry weight)	Squirrel Species	Ingested Energy Assimilated (%)	Ingested Energy Metabolized (%)
White oak, *Quercus alba*	seed kernel	17.5	red	82	78
			gray	77 ± 1.3	72 ± 1.7
			fox	80 ± 0.9	75 ± 1.2
Bur oak, *Q. macrocarpa*	seed kernel	18.2	red	83 ± 0.6	79 ± 0.1
			gray	86 ± 0.9	83 ± 0.7
			fox	85 ± 2.8	81 ± 2.9
Shumard oak, *Q. shumardii*	seed kernel	21.9	red	92	90
			gray	89 ± 1.3	86 ± 1.6
			fox	91 ± 0.3	87 ± 1.1
Black walnut, *Juglans nigra*	seed kernel	26.1	red	94	92
			gray	93 ± 0.04	90 ± 0.4
			fox	93 ± 1.3	90 ± 0.5
Shagbark hickory, *Carya ovata*	seed kernel	27.5	red	95	94
			gray	95 ± 0.7	94 ± 0.8
			fox	95 ± 0.3	94 ± 0.2
Silver maple, *Acer saccharinum*	seed kernel	17.4	red	88 ± 0.1	85 ± 0.4
			gray	82 ± 0.6	80 ± 1.3
			fox	81 ± 1.3	78 ± 0.7
Blackberry, *Rubus ostryifolius*	whole fruit	17.8	red	67 ± 1.7	64 ± 1.4
			gray	62 ± 2.0	59 ± 1.8
			fox	64 ± 1.6	61 ± 1.3
Russula emitica	fungal sporocarp	19.7	gray	61 ± 1.6	53 ± 3.3
Russula sp.	white fungal sporocarp	18.8	gray	67 ± 4.9	57 ± 8.8
			fox	70 ± 3.6	58 ± 6.1
Cantharellus cibarius	fungal sporocarp	19.4	fox	66	51

Source: Smith and Follmer 1972 and unpublished data.
Note: Red squirrel = *Tamiasciurus hudsonicus*; gray squirrel = *Sciurus carolinensis*; fox squirrel = *S. niger*. Values are given with standard errors where available.

lipids. Moreover, the efficiency with which tree squirrels digest the kernels of tree seeds correlates with the lipid content of their foods (table 13.1; Smith and Follmer 1972); this more thorough digestion would further reduce the weight held in the digestive tract when squirrels eat lipid-rich foods. In contrast, ground squirrels eat a high quantity of foliage, which is much lower in lipid content and much higher in relatively indigestible crude fiber. Squirrels of both the genera *Tamiasciurus* and *Sciurus* are able to assimilate from 77% to 95% of the energy in seeds and 61% to 70% in the mushrooms they ingest, whereas the small seeds of blackberry fruits and many of the chitinous cell walls of fungi pass through squirrels undigested, thus reducing the digestive efficiency for these foods (table 13.1; Smith and Follmer 1972). Yellow-bellied marmots (*Marmota flaviventris*) assimilate from 37% to 84% of the energy in foliage they ingest (Kilgore and Armitage 1978); antelope ground squirrels (*Ammospermophilus leucurus*), which feed on both foliage and animal tissues, assimilate from 56% to 68% of the energy in foliage and 85% to 96% of meat or seeds (Karasov 1982); and golden-mantled ground squirrels (*Spermophilus saturatus*), which live in forest edges and eat conifer seeds, fungi, and foliage, assimilate 50% to 52% of foliage and fungi and 96% of conifer seeds (Cork and Kenagy 1989). These variations indicate that the nature of the food, not the type of sciurid, influences the efficiency of digestion.

Nonetheless, the small intestine (165 cm), caecum (12 cm), and large intestine (55 cm) of a 234 g *T. hudsonicus* were 57%, 50%, and 10% longer, respectively, than the equivalent structures in a 263 g Columbian ground squirrel (*Spermophilus columbianus*) (personal observation). The longer, but narrower, small intestine of *T. hudsonicus* is unexpected in relation to the relative digestibility of its typical food but may be adaptive in allowing the emulsification and assimilation of lipids. Moreover, tree squirrels must spend large amounts of time extracting the efficiently digested seed kernels from the hard surroundings in conifer cones and nuts (Smith 1968; Smith and Follmer 1972), and they ingest energy at a slower rate than do ground squirrels. The longer caecum in tree squirrels is also unexpected because researchers believe that they have less crude fiber and cellulose in their diet for microbial digestion in the caecum. The caecum, however, may be involved in the digestion of the chitinous cell walls of fungi, an important part of tree squirrel diets.

Tree squirrels also differ from ground squirrels in their ability to store food for winter. The seeds cached by tree squirrels in the ground are adapted to remain dormant through the winter, and squirrels dry fungi in tree branches so that they will not decompose from bacterial action in storage (Smith and Reichman 1984; Vander Wall and Smith 1987). The foliage eaten by ground squirrels, by contrast, is not naturally dormant and resistant to microbial attack, nor is it easily dried in the open fields and meadows of typical ground squirrel habitat. The silica crystals embedded in mature grass tissue wear down the low-crowned molars of sciurids, so ground squirrels tend to concentrate their feeding on succulent herbs and the soft intercalary meristem of grasses (King 1955; Carleton 1966), which would be particularly difficult to preserve from microbial action. Ground squirrels, therefore, are forced to store body fat and hibernate. Moreover, they cannot synthesize the polyunsaturated fatty acids they need to keep their fat reserves in a liquid and metabolizable state when they arouse from low hibernating temperatures, so they must obtain the fatty acids from plant tissues (Frank 1991, 1992).

The distinction in food resources and habitat between tree and ground squirrels also involves their ways of avoiding predators. As the names imply, ground squirrels retreat to holes they have dug in the ground, whereas tree squirrels run to trees in response to both mammalian and avian predators. Nevertheless, they are similar in that both have short limbs and run with the full surface of their feet on the ground, making them slow on the ground and dependent on alertness to escape. Their forelimbs are powerful in digging and climbing, and the muscles of the forelimb use the same motions in digging and climbing (Stalheim-Smith 1984). The maximum forces exerted by the forelimb muscles of fox squirrels (*Sciurus niger*) are slightly greater and are attained faster than those of Gunnison's prairie dogs (*Cynomys gunnisoni*) of equivalent mass, whereas the muscles of prairie dogs do not fatigue as fast as those of the tree squirrels (Stalheim-Smith 1984). The fact that it requires similar great forearm strength to climb trees and to loosen packed dirt may explain why tree squirrels maintain a light weight by storing almost no body fat even though they have an abundance of food in the trees during the fall or in storage during the winter and why their preferred diet has a very high caloric value. The greater mass of ground squirrels gives an individual more inertia, which serves as an anchor when digging but would be a disadvantage for a climbing tree squirrel.

In addition to having short, strong forelimbs, a high-energy diet, and minimal body fat as adaptations to an arboreal life-style, tree squirrels also have long tails, which they use as a counterbalance when climbing, and unusually large movement in three ankle joints, which allows them to rotate the hind foot 180° when climbing down trees headfirst (Jenkins and McClearn 1984). The sharp, tightly curved claws of tree squirrels allow them to hang by their forefeet when climbing up trees or by their hind feet when climbing down. The longer, broader, and less curved claws of ground squirrels are adapted to the wearing effects of digging.

The Biology of Diurnal Tree Squirrels and Nocturnal Flying Squirrels

Just as the distinction between tree and ground squirrels influences their anatomy, behavior, and ecology, the distinction between diurnal and nocturnal activity patterns creates a marked difference in the lifestyles or niches of arboreal squirrels. Nocturnal tree squirrels are even more specialized to arboreal existence than are diurnal tree squirrels because they have an extended flap of skin (patagium) between the fore and hind limbs that allows them to glide between trees several meters apart with little loss of elevation. These flying squirrels are a monophyletic line (i.e., sharing a common evolutionary ancestry) derived over 17 million years ago from the diurnal tree squirrels (Thorington 1984).

In the temperate deciduous forests of North America, southern flying squirrels (*Glaucomys volans*) eat and cache nuts from the dominant tree species in the same general manner as do sympatric species of *Sciurus* but have a greater tendency to cache food in tree cavities than do diurnal tree squirrels (Muul 1968). In boreal forests, by contrast, northern flying squirrels (*G. sabrinus*) eat fungi and foliose arboreal lichens (McKeever 1960; Maser et al. 1978) and have not been reported to cache food. At the high latitudes of boreal forests, summer days and winter nights make up the major portion of 24-hour periods. Animals should be selected to use days or nights dependent on which is longest during the part of the year when the animals are most limited for time: late summer for caching food for diurnal *Tamiasciurus* and winter for foraging for nocturnal *Glaucomys*. The same distinction for activity patterns applies between lagomorphs where diurnal pikas (*Ochotona*) cache food in late summer and crepuscular to nocturnal snowshoe hares (*Lepus americanus*) must forage for food all winter.

These two types of squirrels also differ in their vision. Diurnal tree squirrels, like ground squirrels, have pure cone retinas and apparently distinguish colors, whereas nocturnal flying squirrels have pure rod retinas and cannot see color (Young and Hobbs 1975). One might expect that the fields of vision of the two eyes of all arboreal squirrels would overlap extensively to allow binocular vision for judging distances when jumping between trees; however, the eyes of diurnal tree squirrels are relatively far back on the sides of the face, just like those of ground squirrels. Diurnal tree squirrels are quite cautious in making what appear to be short jumps. By contrast, the eyes of flying squirrels are positioned much farther forward, and researchers have used the presence of a central circular blind spot in their retinas (Young and Hobbs 1975) to argue that the fields of vision of their two eyes do overlap extensively. In diurnal tree and ground squirrels, the optic tract leaves the eye in a dorsally positioned horizontal line so that the blind spot will not interfere with the vertical discrimination of enemies approaching from above. Flying squirrels must be able to judge distances to land in trees in a coordinated manner after a long glide. The ability to glide would appear to be a necessity for nocturnal tree squirrels, all of which are volant throughout the world. Gliding allows the squirrels to escape predators (e.g., marten, *Martes americana*) that are able to approach closely within trees in the dim light of night.

Temperature Regulation in Diurnal Tree Squirrels

Even the most northern of diurnal tree squirrels have relatively thin fur and a lower critical temperature of 11°C in an Alaskan winter (Irving et al. 1955). The presence of stored food is probably the factor that allows these squirrels to have such a relatively high value for their lower critical temperature (i.e., the environmental temperature below which an animal must expend extra energy to maintain a constant normal body temperature). With stored food, squirrels can minimize their foraging time and use their nest for insulation during inclement weather. Tree squirrels bask to warm their body temperature when leaving the nest on cold sunny days (Golightly and Ohmart 1978; Smith 1981). The percentage of agouti guard hairs that is made up of black bands in *Tamiasciurus* increases in winter (Smith 1981), and the frequency of melanic (black) morphs (or forms) increases at higher latitudes in three species of *Sciurus*. Black fur may increase the conversion of light to heat in basking squirrels in colder climates; however, researchers have found that although black morphs of *S. carolinensis* had lower costs of thermoregulation than gray morphs, the thermal conductance of the two forms did not differ (Innes and Lavigne 1979). Whatever the thermal function of black in guard hairs, basking as a form of conserving metabolic energy is potentially more useful in species that store food and have more flexible time budgets for feeding.

Contrasting the Biology of *Tamiasciurus* and *Sciurus*

In North America the two genera of diurnal tree squirrels are largely separated, with *Tamiasciurus* in the boreal and montane conifer forests and the Pacific Coastal conifer forests, while *Sciurus* is located in temperate deciduous forests and in lower-elevation and more southern pine forests. Where the boreal and temperate deciduous forest biomes mix, the two genera of squirrels segregate by habitat with intergeneric aggression having little to do with the spacing (Riege 1991).

To a large extent the differences in the biology of the two genera are based on radical differences in the seed-producing structures of the trees in boreal and temperate deciduous forests. In boreal forests both conifers and angiosperms have many seeds packaged together in female catkins, whereas the angiosperms of the temperate deciduous forests, with very few exceptions, have single-seeded fruits. Some of my coworkers and I have attributed this difference to the effect of tree species diversity on the frequency of successful cross-pollination by wind in the two major biomes (Smith et al. 1990). In boreal forests, which have low species diversity, a high frequency of females will receive conspecific cross-pollen (i.e., from a genetically different tree) and the trees commit energy to woody protective tissue in female catkins (cones) over a wide range of seed set. Because many seeds are packaged in a catkin, the extensive energy commitment to woody protective tissue is independent of the frequency of seed set. Mast fruiting (synchronous high variation in annual seed crops) in this biome is a consequence of mast flowering of both sexes so that most female flowers are produced during years of abundant pollen. In contrast, in temperate deciduous forests, which have higher species diversity, individual trees are likely to experience a higher frequency of self pollen among conspecific pollen and, if they are dioecious (male and female flowers in different trees), to experience less conspecific pollen than trees in boreal forest. Trees in the temperate deciduous forests with single-seeded fruits commit very little energy to female function until after fertilization. The energy thus saved would be lost if many seeds were mutually defended in a common fruiting body. Mast fruiting of acorns, beech nuts, and other similar seeds in the temperate deciduous forest biome is therefore not a consequence of mast flowering but rather the result of differential female mortality after nearly uniform annual flower production. In both biomes, most species of trees experience large annual variation in seed crop (Smith 1968, 1970; Janzen 1971; Silvertown 1980), and squirrels in both the genera *Sciurus* and *Tamiasciurus* respond to crop failures by emigrating in large numbers from the affected habitat (Schorger 1949; Smith 1968, 1970).

The difference in the structure of fruiting bodies between boreal and temperate trees results in a difference in the ability of squirrels to defend stored food against other species of seed eaters (Stapanian and Smith 1978). *Tamiasciurus* stores whole conifer cones in large concentrated caches (middens) in damp places in the ground, which prevents the cones from drying and opening their scales to release seeds. The woody scales and bracts of conifer cones are probably too hard for mice to open and obtain seeds at a rate faster than they can find individual seeds on the ground. I have found no evidence of mice gnawing at cones in squirrel caches, which they could certainly do without threat from squirrels at night (Smith 1968). The high percentage (> 70%) of woody tissue in most species of conifer cones (Smith 1970) should preclude larger animals, such as deer and bear, from eating whole conifer cones from squirrel caches to obtain the digestible energy in seeds. Thus, cached, closed conifer cones apparently cannot be used efficiently by other mammals or birds with three possible exceptions. First, grizzly bears (*Ursus arctos*) and black bears (*U. americanus*) chew or tear open cones of three large-seeded pine species in squirrel caches and lick up the released pine nuts (Kendall 1983). These pines are species whose seeds are dispersed by nutcrackers (*Nucifraga* sp.) to dry, high-altitude habitats (Vander Wall 1990) of rather limited area compared to the total range of *Tamiasciurus*. Second, the relatively large Townsend's chipmunks (*Eutamias townsendi*) in the Pacific coastal conifer forests chew open closed Douglas-fir (*Pseudotsuga menziesii*) and western hemlock (*Tsuga heterophylla*) cones and are chased away from caches by *Tamiasciurus* (Smith 1968). The chipmunks hibernate in winter and should not pose a significant threat to squirrel caches. Third, Abert's squirrels (*Sciurus aberti*) take Douglas-fir and perhaps other conifer cones from *Tamiasciurus* caches in the Pinaleño Mountains (Paul Young, personal communication). Abert's squirrels may also take cones from caches when in contact with *Tamiasciurus* in other parts of its range, but other researchers have reported no evidence of this interaction.

In contrast to the lack of interspecific competitors for most concentrated caches (also called larder hoards or middens) of *Tamiasciurus*, squirrels in the genus *Sciurus* have several species of birds and mammals larger than themselves that may effectively compete for larder hoards of acorns and beech and hickory nuts. The ratio of digestible kernel to indigestible shell is high enough in these nuts to allow deer (*Odocoileus*), black bears, and turkeys (*Meleagris gallopavo*) to eat whole nuts. Moreover, mice (*Peromyscus*), opossums (*Didelphis virginiana*), and raccoons (*Procyon lotor*) could use the *Sciurus* caches at night, and several species of smaller birds could fly in quickly and raid the caches during the day. This inability to defend larder hoards probably selects for the pattern of scatter-hoarding nuts found in *Sciurus* (Stapanian and Smith 1978, 1984). Individual squirrels bury individual nuts spaced in the area around their den tree such that the difference between the cost of burying nuts and the benefit of being able to retrieve the nuts at a later date is maximized (Stapanian and Smith 1978). Burying nuts eliminates their use by birds, which do not have a good enough sense of smell to detect them under soil, and optimum spacing minimizes the probability that squirrels naive to their burial location will find them by density-dependent search after chancing to find one by its odor.

The difference in the packaging of seeds in boreal and temperate forests and the difference in pattern of storage they allow squirrels to use have in turn selected for radical differences in the social behavior of *Tamiasciurus* and *Sciurus*. *Tamiasciurus* has evolved a social system in which individuals, either male or female, defend territories (figure 13.1) that allow individual squirrels to minimize the distance traveled and the time and energy expended in cutting cones from trees and concentrating them in a larder hoard near the center of a territory (Smith 1968). Territory area is thus inversely proportional to food density (Smith 1968). After cutting and caching cones in the fall, individuals feed from their larder during the winter. Caches are usually around large fallen trees so that the squirrels can feed under the snow during poor weather. Because *Tamiasciurus* squirrels exhibit no intergeneric competition for cached cones in most situations, the territorial system fits Jerram Brown's (1964) requirements of a defensible resource in short supply as the basis for the evolution of territorial systems.

Tamiasciurus scatter-hoards fungi in the upper branches of trees

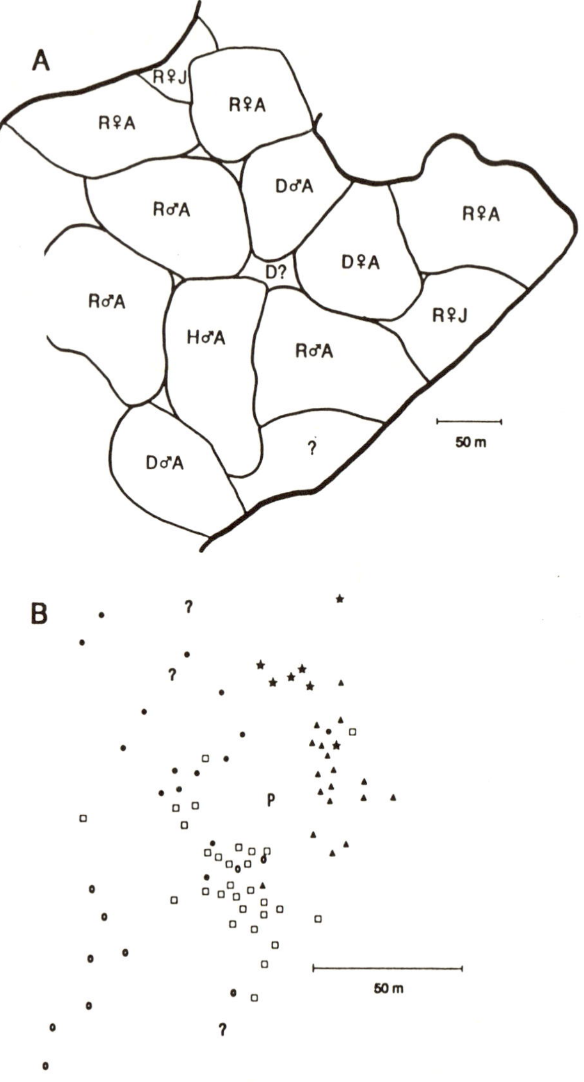

Figure 13.1 Territories of 13 individual *Tamiasciurus* in a lodgepole pine forest (Figure 13.1a) contrasted with the spacing of black walnuts (*Juglans nigra*) cached by five individual *Sciurus niger* in a wooded cemetery (Figure 13.1b). Figure 13.1a: Territory boundaries enclose positions in which individuals used territorial calls. D = *T. douglasii*, R = *T. hudsonicus*, and H = a hybrid individual; A (inside a territory) = adult, and J = juvenile (Smith 1968). The unsexed *T. douglasii* was just starting to call from an undefended area between five other territories when I left

to dry (Smith 1968). The fungi are usually left scattered, but in the two instances in which they have been reported grouped (Buller 1920; Smith 1968), they were in the active nest of a squirrel, the only place they could be defended at night from fungus-feeding northern flying squirrels.

The territorial system may break down completely for two or three weeks in the spring when squirrels feed together on superabundant but nonstorable conifer pollen (Smith 1968). The social behavior of *Tamiasciurus* that accompanies this territorial system is communicated by five distinct vocalizations (Smith 1978; Lair 1990): (1) a loud rattle call of rapidly repeated notes used by territory owners within the boundaries of their territories to advertise their ownership and by males towards each other during mating chases; (2) a loud screeching call made of drawn-out rattle call notes that territory owners use when they detect another squirrel entering their territory; (3) a quiet growling call used by squirrels in a fight or in a situation that is likely to lead to a fight; (4) a quiet buzzing call used in situations where one squirrel approaches another without aggression, such as males approaching females in estrus, mothers approaching their young to nurse them, or juveniles approaching adults after the juveniles first leave the nest; and (5) a chirping call of more widely spaced notes used to warn other squirrels of the presence of potential predators. The short dispersal distances of juvenile *Tamiasciurus* (Larsen and Boutin 1994) make it likely that the apparent altruism of chirping calls is based on kin selection.

By contrast, the scatter-hoarded caches of individual *Sciurus* (figure 13.1) cannot be effectively observed and defended from other squirrels by a territorial system (Stapanian and Smith 1978). Squirrels in this genus have broadly overlapping home ranges (Gurnell 1987), and they lack the rattle and screeching calls that individual *Tamiasciurus* use to advertise and defend their territories. *Sciurus* squirrels do have calls that are equivalent in function to the other three *Tamiasciurus* calls

---

the area, and the question mark represents an undefended area of more open forests. Heavier lines show the border of the forest. Figure 13.1b: The position of each of the buried walnuts is denoted by a different symbol for each of the five individual *S. niger* (Stapanian and Smith 1978). All nuts were taken from a central pile (P), which we supplied after stripping the trees of walnuts. Nuts buried by a squirrel of unknown individual identity are denoted by a question mark.

(Gurnell 1987), and the mating behavior of the two genera is similar, with several males giving buzzing calls while trying to approach a female during her one day of estrus (Bakken 1959; Smith 1968; Thompson 1977; Koford 1982; Gurnell 1987; Koprowski 1993). A dominant male follows less than a meter behind a female in estrus and drives off the other males until, after as much as four hours, the female accepts copulation.

Although the behavior of the two genera is much the same on the day of estrus, the territorial system of *Tamiasciurus* makes the coordination of the sexes different. The vagina of a female *Tamiasciurus* approaching estrus is much more enlarged and convoluted (Mossman et al. 1932) than that of a *Sciurus* female, apparently to produce a secretion that is left around her territory to communicate her reproductive condition to males who wander far from their own territories during the mating season (Smith 1968). In *Sciurus,* males can approach females directly to determine their reproductive condition. The convoluted vagina in *Tamiasciurus* matches the unusually long penis of the male, which, unlike the *Sciurus* male, lacks a baculum (penis bone) (Mossman et al. 1932). Also unlike *Sciurus,* the distal male accessory glands (Cowper's and bulbo-urethral) of *Tamiasciurus* are greatly reduced and the proximal seminal vesicles much enlarged, perhaps to free the long penis for penetrating the long convoluted vagina of the female (Mossman et al. 1932).

All of these differences in the reproductive tracts of *Sciurus* and *Tamiasciurus* could well result from the advantage to female *Tamiasciurus* of signaling their reproductive status while simultaneously defending a territory against males and females except for that one day of estrus. Males would stand little chance of losing their territories while involved in mating behavior during the late winter and early spring because squirrels who were unable to obtain a territory during the previous fall would likely have starved during the winter (Larsen and Boutin 1994). Males might lose much of their remaining caches to neighboring female territory owners during the mating season, although I know of no observation of such losses.

Various species of *Sciurus* rather than *Tamiasciurus* exploit the lower-elevation pine forests of the southeastern and southwestern United States and Central America (Hall and Kelson 1959; Farentinos 1972). These species of *Sciurus* chew open pine cones to extract seeds, but they do not cut unopened cones and cache them. In contrast to *Tamiasciurus,* the absence of either larder-hoarding or scatter-hoarding of either whole cones or seeds by *Sciurus* invites an explanation. The warmer temperatures, drier soils, and longer snow-free period in the fall in lower-elevation and lower-latitude pine forests probably prevent the economic caching of cones in a condition that would keep them closed and their seeds free from theft by nocturnal mice. The high ratio of woody protective tissue to digestible seed kernels that makes it uneconomical for mice to open closed cones also makes it uneconomical for squirrels to dig holes deep enough to bury whole cones in an environment damp enough to keep them closed. In the damper, high-elevation forests where *Tamiasciurus* larder-hoards cones, squirrels often pile many cones together on the surface of the ground and spend relatively little time burying the cones that are left in rotting dissected cone pieces of old middens. It is unlikely that *Sciurus* could economically cache pine seeds outside of cones. A squirrel is too large to compete with mice and birds for individual seeds that have fallen to the ground from open cones. If the squirrels extract the seeds from closed cones before caching them, the rate at which they could store their winter food supply would be greatly slowed, and they would have to scatter their caches to minimize loss to mice, which could take the seeds unprotected by cone tissue. Various species of birds in the family Corvidae are much more efficient than slower-moving squirrels could be at this means of caching pine seeds (Vander Wall 1990).

Among the species of *Sciurus* using pine seeds, the Abert's squirrel of ponderosa pine (*Pinus ponderosa*) forests of the southwestern United States and northern Mexico is the most dependent on pines. But even this species does not cache cones or seeds, switching instead to eating bark from terminal twigs once the cones shed their seeds (Farentinos 1972).

At the upper elevations of the ponderosa pine zone in the Cascade Mountains of Washington and British Columbia, *Tamiasciurus* caches cones of ponderosa pine along with Douglas-fir in damp middens, but middens are absent in drier open stands at lower elevations (personal observation). It would appear that damp areas for caching are critical to their evolution of larder-hoarding and the territorial defense of caches.

In the forests of Eurasia *Sciurus vulgaris* seems to exhibit the caching behavior and social organization typical of the genus in spite of being found in damp boreal forests as well as temperate deciduous forests (Gurnell 1987). In England, where *S. vulgaris* competes with *S. carolinensis* introduced from the temperate deciduous forests of North

America, *S. vulgaris* tends to be displaced from the deciduous forests and is able to maintain itself mainly in the northern conifer forests. If *Tamiasciurus* were also introduced into England, its pattern of larder-hoarding conifer cones would probably give it an advantage over *S. vulgaris* in the conifer forests, driving the native squirrel to extinction. *S. vulgaris* would thus appear to be an example of the jack of all trades and master of none.

Tamiasciurus hudsonicus grahamensis in Comparison to Other Populations of *Tamiasciurus*

Tamiasciurus hudsonicus grahamensis, the Mt. Graham red squirrel, is isolated in the higher elevations of the Pinaleño Mountains in southeastern Arizona. It probably has not had genetic contact with other populations of *Tamiasciurus* since the last period of Pleistocene glaciation 10,000 to 12,000 years ago (Chapter 12; Lomolino et al. 1989). Researchers have found that both emigration through low-elevation oak-pine woodlands and the area of montane forest islands together give the best explanation of species richness in montane forest mammal communities (Chapters 9, 11; Lomolino et al. 1989). The Pinaleño Mountains, however, are isolated by over 30 km of chaparral, grassland, or desert scrub, which *Tamiasciurus* would be unlikely to cross.

In the Pinaleños, red squirrels are concentrated in spruce-fir forests (*Picea engelmannii–Abies lasiocarpa*) at the highest elevations and in mixed-conifer forests composed largely of Douglas-fir and white fir (*A. concolor*) just below. The presence of large middens of old, rotting, dissected cone parts and caches of newly stored whole conifer cones along with observations of feeding on fungal sporocarps indicate that the biology of this subspecies is typical of the genus. However, it differs from other populations that I have observed in the lower density of middens and the lower frequency of calling by individual squirrels. Other researchers have extensively documented the density of middens (Froehlich 1990) and have confirmed my impression of a low frequency of calling (J. H. Brown and P. Young, personal communication). These traits could be related to the small area of prime habitat in the mountains and the lack of alternative habitat into which the squirrels could emigrate during cone failures in the four species of conifers upon which they depend. During the long period of isolation there may have been natural selection for larger territories and reduced local migration to allow squirrels to survive low cone crops without emigration. The paucity of emigrating squirrels would reduce the selective advantage of frequent calling to advertise territorial occupancy. If squirrel calls also alert predators to the location of the caller, there should have been a selective pressure to reduce the frequency of calling.

The logging of some preferred habitat over the last century in the Pinaleños and the introduction of Abert's squirrels in the 1940s have no doubt reduced the size of the red squirrel population, but the relatively restricted area on the top of the mountains where conifer cones could be cached in autumn soils moist enough to preserve a winter food supply could never have supported more than a few thousand squirrels when all the forests were in a mature condition. During periodic failures in the cone crop, the squirrel populations would reach bottlenecks in size that would have made this a threatened species for 10,000 years. Thus, this population is particularly valuable for study as an example of what factors will allow a population to survive an extended period at a small population size.

References

Bakken, A. 1959. Behavior of gray squirrels. *Proceedings of the Annual Conference of the Southeastern Association of Game and Fish Commissioners* 13:393–406.

Brown, J. L. 1964. The evolution of diversity in avian territorial systems. *Wilson Bulletin* 76:160–69.

Buller, A. H. R. 1920. The red squirrel of North America as a mycophagist. *Transactions of the British Mycological Society* 6:355–62.

Carleton, W. M. 1966. Food habits of two sympatric Colorado sciurids. *Journal of Mammalogy* 47:91–103.

Cork, S. J., and G. J. Kenagy. 1989. Nutritional value of hypogeous fungus for a forest-dwelling ground squirrel. *Ecology* 70:577–86.

Farentinos, R. C. 1972. Observations on the ecology of the tassel-eared squirrel. *Journal of Wildlife Management* 36:1234–39.

Frank, C. L. 1991. Adaptations for hibernation in the depot fats of a ground squirrel (*Spermophilus beldingi*). *Canadian Journal of Zoology* 69:2707–11.

Frank, C. L. 1992. The influence of dietary fatty acids on hibernation by golden-mantled ground squirrels (*Spermophilus lateralis*). *Physiological Zoology* 65:906–20.

Froehlich, G. F. 1990. Habitat use and life history of the Mt. Graham red squirrel. M.S. thesis, University of Arizona, Tucson.

Golightly, R. T., Jr., and R. D. Ohmart. 1978. Heterothermy in free-ranging Abert's squirrels (*Sciurus aberti*). *Ecology* 59:897–909.

Gurnell, J. 1987. *The Natural History of Squirrels*. Facts on File Publishers, New York.

Hall, E. R., and K. R. Kelson. 1959. *The Mammals of North America*. Ronald Press, New York.

Innes, S., and D. M. Lavigne. 1979. Comparative energetics of coat colour polymorphisms in the eastern gray squirrel, *Sciurus carolinensis*. *Canadian Journal of Zoology* 57:585–92.

Irving, L., H. Krog, and M. Monson. 1955. The metabolism of some Alaskan animals in winter and summer. *Physiological Zoology* 28:173–85.

Janzen, D. H. 1971. Seed predation by animals. *Annual Review of Ecology and Systematics* 2:465–92.

Jenkins, F. A., Jr., and D. McClearn. 1981. Mechanisms of hind foot reversal in climbing mammals. *Journal of Morphology* 182:197–219.

Karasov, W. H. 1982. Energy assimilation, nitrogen requirement, and diet in free-living antelope ground squirrels *Ammospermophilus leucurus*. *Physiological Zoology* 55:378–92.

Kendall, K. C. 1983. Use of pine nuts by grizzly bears in the Yellowstone area. *International Conference of Bear Research and Management* 5:166–73.

Kilgore, D. L., and K. B. Armitage. 1978. Energetics of yellow-bellied marmot populations. *Ecology* 59:78–88.

King, J. A. 1955. Social behavior, social organization, and population dynamics in a black-tailed prairie dog town in the Black Hills of South Dakota. *University of Michigan Contributions of the Laboratory of Vertebrate Biology* 67:1–123.

Koford, R. R. 1982. Mating system of a territorial tree squirrel (*Tamiasciurus douglasii*) in California. *Journal of Mammalogy* 63:274–83.

Koprowski, J. L. 1993. Alternative reproductive tactics in male eastern gray squirrels: "Making the best of a bad job." *Behavioral Ecology* 4:165–71.

Lair, H. 1990. The calls of the red squirrel: A contextual analysis of function. *Behavior* 115:254–82.

Larsen, K. W., and S. Boutin. 1994. Movement, survival, and settlement of young red squirrels. *Ecology* 75:214–23.

Lomolino, M. V., J. H. Brown, and R. Davis. 1989. Island biogeography of montane forest mammals in the American Southwest. *Ecology* 70:180–94.

McKeever, S. 1960. Food of the northern flying squirrel in northeastern California. *Journal of Mammalogy* 41:270–71.

Maser, C., J. M. Trappe, and R. A. Nussbaum. 1978. Fungal–small mammal interrelationships with emphasis on Oregon coniferous forests. *Ecology* 59: 799–809.

Mossman, H. W., J. W. Lawlor, and J. A. Bradley. 1932. The male reproductive tract of the Sciuridae. *American Journal of Anatomy* 51:89–141.

Muul, I. 1968. Behavioral and physiological influences on the distribution of the flying squirrel, *Glaucomys volans*. *Miscellaneous Publications, Museum of Zoology, University of Michigan* 134:1–66.

Riege, D. A. 1991. Habitat specialization and social factors in distribution of red and gray squirrels. *Journal of Mammalogy* 72:152–62.

Schorger, A. W. 1949. Squirrels in early Wisconsin. *Transactions of the Wisconsin Academy of Science, Arts, and Letters* 39:195–247.

Silvertown, J. W. 1980. The evolutionary ecology of mast seeding in trees. *Biological Journal of the Linnean Society* 14:235–50.

Smith, C. C. 1968. The adaptive nature of social organization in the genus of tree squirrel *Tamiasciurus*. *Ecological Monographs* 38:30–63.

Smith, C. C. 1970. The coevolution of pine squirrels (*Tamiasciurus*) and conifers. *Ecological Monographs* 40:349–71.

Smith, C. C. 1978. The structure and function of the vocalizations of tree squirrels (*Tamiasciurus*). *Journal of Mammalogy* 59:793–808.

Smith, C. C. 1981. The indivisible niche of *Tamiasciurus*, an example of non-partitioning of resources. *Ecological Monographs* 51:343–63.

Smith, C. C., and D. Follmer. 1972. Food preferences of squirrels. *Ecology* 53:82–91.

Smith, C. C., J. L. Hamrick, and C. L. Kramer. 1990. The advantage of mast years for wind pollination. *American Naturalist* 136:154–66.

Smith, C. C., and O. J. Reichman. 1984. The evolution of food caching by birds and mammals. *Annual Review of Ecology and Systematics* 15:329–51.

Stalheim-Smith, A. 1984. Comparative study of the forelimbs of the semi-fossorial prairie dog, *Cynomys gunnisoni*, and the scansorial fox squirrel, *Sciurus niger*. *Journal of Morphology* 180:55–68.

Stapanian, M. A., and C. C. Smith. 1978. A model for seed scatter hoarding: Coevolution of fox squirrels and black walnuts. *Ecology* 59:884–96.

Stapanian, M. A., and C. C. Smith. 1984. Density-dependent survival of scatter hoarded nuts: An experimental approach. *Ecology* 65:1387–96.

Thompson, D. C. 1977. Reproductive behaviour of the gray squirrel. *Canadian Journal of Zoology* 55:1176–84.

Thorington, R. W., Jr. 1984. Flying squirrels are monophyletic. *Science* 225: 1048–50.

Vander Wall, S. B. 1990. *Food Hoarding in Animals*. University of Chicago Press, Chicago.

Vander Wall, S. B., and K. G. Smith. 1987. Cache-protecting behavior of food-hoarding animals. In *Foraging Behavior*, ed. A. C. Kamil, J. R. Krebs, and R. H. Pulliam, pp. 611–44. Plenum Press, New York.

Young, J. Z., and M. J. Hobbs. 1975. *The Life of Mammals*. Clarendon Press, Oxford, U.K.

CHAPTER FOURTEEN

Monitoring the Mt. Graham Red Squirrel

Paul J. Young

In 1988, Congress passed the Arizona-Idaho Conservation Act. One section of the act directed the Forest Service to issue permits for the construction of the Mt. Graham International Observatory (MGIO), initially with three telescopes, and possibly four more later on, in the Pinaleño Mountains of southeastern Arizona (see Chapters 1–3). The permit included many conditions and mitigating actions because the observatory was to be constructed in the middle of the habitat of the endangered Mt. Graham red squirrel (*Tamiasciurus hudsonicus grahamensis*); the red squirrel had been listed as a federally endangered subspecies in 1987.

The permit required that the University of Arizona, one of the sponsors of the observatory, fund a program to monitor the red squirrel population around the observatory site. The researchers conducting the monitoring program were to collect data on the impact of the construction and operation of the observatory on the red squirrel population and were to report these data to the Forest Service. Local administrators of the Forest Service were to examine the findings of the monitoring program to determine whether the effects of the observatory on the Mt. Graham red squirrel population exceeded those predicted and allowed by the 1988 biological opinion issued by the Fish and Wildlife Service (Spear 1988) that had set the conditions for building the telescopes. Additionally, the resource managers were to apply the data and findings of the monitoring program to future decisions about the possible construction of additional telescopes.

With the exception of a few other independent studies (Spicer et al. 1985; Froehlich 1990; Smith and Mannan 1994), the monitoring program has been virtually the sole source of information on the life history and population biology of the Mt. Graham red squirrel. Our studies have begun to answer some of the many questions concerning the ecology of this endangered subspecies. The university and its partners in the observatory are required to maintain the monitoring program for the life of the observatory; this requirement presents a unique opportunity for a long-term ecological study of the red squirrel.

At the insistence of the overseeing government agencies (especially the Forest Service and Fish and Wildlife Service), the original monitoring plan (Coronado 1989) required behavioral observations rather than a focus on population numbers. However, behavioral observations proved to be of little value for assessing the impact of the observatory, partly because very few squirrels lived close to the construction and partly because the behavior of individual squirrels was so variable that enormous amounts of observation time were required to discriminate between differences in behavior patterns of squirrels close to construction and those on control areas. We discontinued most behavioral studies after 1991 and have since concentrated on documenting changes in the number and distribution of squirrels.

We have attempted to capture and individually mark squirrels to document their movements and to improve life history data, but these attempts have had minimal success. Trapping efforts from 1989 through 1992 resulted in an average of about 100 trap-days per capture, and an unusually high number of trapping-related squirrel deaths (3 of 28 captures) have ensued. Additionally, under the permit issued by the Fish and Wildlife Service, biologists conducting the monitoring program are not allowed to trap and mark juvenile squirrels, making it impossible to obtain accurate data on life spans and dispersal patterns. Trapping efforts have been further complicated by the recent outbreak of a hantavirus disease in Arizona. Hanta or hantalike viruses have not been isolated from red squirrels to date, but the fact that the disease can pass from rodents to humans, suggesting it can also pass between rodent species, is sufficient cause for concern. Therefore, we had to take safety precautions (wearing gloves, respirators, and protective clothing and disinfecting traps after each use) that made the trapping and handling of squirrels a cumbersome, nearly impossible, task under the conditions and rugged terrain of the monitored areas.

Nevertheless, the monitoring program from August 1989 through June 1994 was productive. We documented changes in the red squirrel

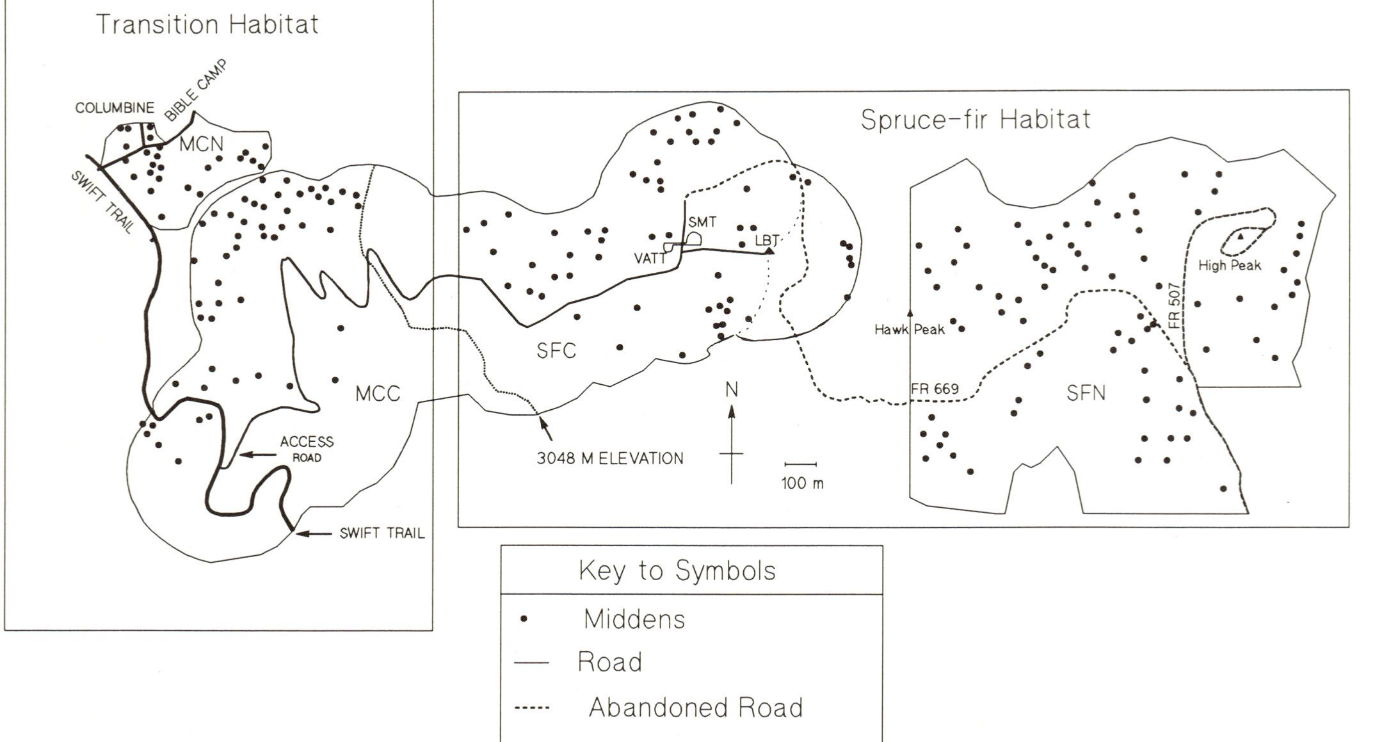

Figure 14.1 Map of the areas in the Pinaleño Mountains monitored by the University of Arizona Mt. Graham red squirrel monitoring program. Within the transition, or mixed-conifer, habitat is a large mixed-conifer construction study area (MCC) and a small adjacent mixed-conifer nonconstruction study area (MCN). The corresponding study areas in the spruce-fir habitat are the spruce-fir construction area (SFC) and the spruce-fir nonconstruction area (SFN). The dashed portions of forest roads 669 and 507 (FR 669, FR 507) mark the roads removed as a mitigating measure when observatory construction began. VATT is the Vatican Advanced Technology Telescope, SMT is the Submillimeter Telescope, and LBT is the Large Binocular Telescope.

populations on the areas monitored by the university, and these changes indicate the overall response of the population of this endangered subspecies to the construction of the MGIO.

Description of the Monitored Areas

The MGIO is being built on Emerald Peak in the Pinaleño Mountains of southeastern Arizona (see figure 5.4). The construction area, for squirrel monitoring purposes, was defined in the MGIO management plan (Coronado 1989) as all the area within 1,000 feet (304.7 m) of the proposed telescope sites and the access road (figure 14.1); the boundary was changed to 300 m in 1989. The 3,050 m (10,000 foot) elevation contour approximates the boundary between two high-elevation coniferous forest habitat types found on the Pinaleño Mountains—montane coniferous (or mixed-conifer) forest and subalpine coniferous (or spruce-fir) forest (Chapter 6; Chase and Brown 1982a, 1982b)—and separates the construction area into two areas of almost equal size.

The mixed-conifer habitat construction area (MCC) constitutes the western portion of the construction area, below 3,050 m elevation (about 91 ha: transition habitat, figure 14.1). The forest cover on the MCC area varies in species composition and dominance. The primary tree species in this area are Douglas-fir (*Pseudotsuga menziesii*), corkbark fir (*Abies lasiocarpa* var. *arizonica*, now *A. bifolia*), Engelmann spruce (*Picea engelmannii*), ponderosa pine (*Pinus ponderosa*), Western white pine (*Pinus strobiformis*), and aspen (*Populus tremuloides*).

The spruce-fir habitat construction area (SFC) is the eastern portion of the construction area, above 3,050 m elevation (originally about 89 ha), and is primarily spruce-fir forest, dominated by Engelmann spruce and corkbark fir. The monitoring program added 14 ha to the SFC area in late 1994 when the Forest Service allowed the MGIO astronomers to locate one of the first three telescopes farther east than was originally planned.

For comparison, we are also monitoring squirrel populations on two nonconstruction areas of the Pinaleño Mountains as a control population (figure 14.1). We originally chose a spruce-fir nonconstruction area (SFN) about 0.75 km east of the east boundary of the SFC area as the only nonconstruction area (figure 14.1). The west boundary of this area was defined as a line running from the top of Hawk Peak to an elevation of 3,050 m on the south and to a hiking trail on the north. The north boundary follows this hiking trail eastward to a point north of

High Peak, then crosses over the ridge on the northeastern corner of that peak and continues down the eastern side of the area along the 3,050 m contour. The south boundary follows the 3,050 m elevation contour to a point about 1 km south of High Peak, then crosses over Forest Road 507 (FR 507) to connect to the east boundary. This SFN has an area of about 122 ha.

In September 1989, we added a small area (about 22 ha) of mixed conifer forest habitat northwest of the MCC area to compare the squirrel population on the MCC area with a population in a similar habitat (Young 1991). The new area, referred to as the mixed-conifer habitat nonconstruction area (MCN), is adjacent to the northwest boundary of the MCC area and is bordered on the west by state highway 366, the Swift Trail (figure 14.1). We arbitrarily determined the borders of the MCN area by drawing a line about 100 m to the outside of the known midden sites. The size of this area was increased to 25 ha in 1990 to include three new middens north of the Bible camp road.

Census Methods

The characteristics of squirrel habitat on the mountain range vary considerably from area to area (Coronado 1988), and the four monitored areas differ in size, forest species composition, and the amount of forest cover on each area. The total number of squirrels on each area, therefore, is not a sufficient comparison among populations. To ensure valid assessments of the effect of construction on squirrel populations the variables measured must be as independent of the size and habitat quality of the areas as possible. Nevertheless, an accurate count of the number of red squirrels on each area provides the starting basis for a comparison of populations in construction and nonconstruction areas.

Data on the number and distribution of squirrels are collected during a regular census of each area. In the snow-free period of the year a census is conducted each month; in winter and spring, when snow covers most of the monitored areas to depths up to 3 m, a census is conducted every other month. We visit each known midden site at least once during each census to determine its occupancy status. If a midden appears to be occupied, on the basis of feeding sign or caching, we attempt to observe the squirrel and to determine its gender, age class, and reproductive condition. During winter months, sightings of squirrels often are impractical, and determinations of occupancy are based on the presence and age of feeding sign, tracks, and other indications of red squirrel activity. A midden is considered unoccupied when no squirrel or squirrel sign is observed or the status of the midden is uncertain. Since red squirrels are highly territorial, and generally only one squirrel occupies a midden (Chapter 13; Vahle 1978), we consider, when determining the size of the population, that an occupied midden represents one squirrel. All analyses are conducted between construction and nonconstruction areas within habitats, because data for the behavior, numbers, and densities of squirrels for populations in mixed-conifer versus spruce-fir habitats are not comparable.

To compare squirrel populations among different areas and habitats in an objective manner, one less subject to variation in the size, quality, or patchiness of the areas, we use indices of population size and distribution. The index used for the size of the population is the proportion of known, occupied midden sites on each area. The hypothesis to be tested is that construction activity close to a midden would result in making that site less suitable for occupancy; therefore we analyze midden occupancy for statistical differences between construction and nonconstruction areas within each habitat using the data from the June and November (or December) censuses of each year. Within the construction areas, we also analyze midden occupancy relative to distance from construction using data from the last census of the year.

We compare densities of squirrels and midden sites, as opposed to total numbers, using an index of density that we have termed "local density." Local density is similar in concept to distance methods of neighborhood modeling used in plant ecology (Czaran and Bartha 1992). The local density of middens is the average number of middens found in a circle 100 m in radius around each midden site. Similarly, the local density of squirrels is the average number of squirrels found within 100 m of each squirrel.

During, and between, each regular census we observe reproductive activity, the reproductive status of individuals, the mating success of females (whether or not they produce a litter), and the litter size. We visit lactating females every few days to locate the natal nest and to count the number of offspring after emergence. We then compare the proportion of lactating females per area, the proportion of lactating females that successfully rear litters, and the litter size on each area between construction and nonconstruction areas and between habitats. An analysis and summary of the data collected during each year is presented to the Forest Service (Safford District Ranger) in an annual report. The final decision on whether construction and operation of the

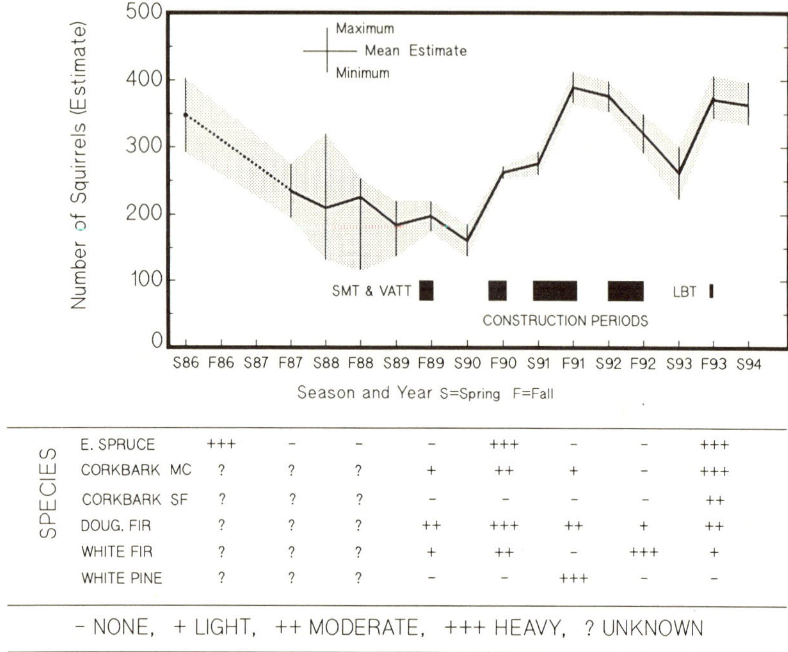

SPECIES								
E. SPRUCE	+++	-	-	-	+++	-	-	+++
CORKBARK MC	?	?	?	+	++	+	-	+++
CORKBARK SF	?	?	?	-	-	-	-	++
DOUG. FIR	?	?	?	++	+++	++	+	++
WHITE FIR	?	?	?	+	++	-	+++	+
WHITE PINE	?	?	?	-	-	+++	-	-

- NONE, + LIGHT, ++ MODERATE, +++ HEAVY, ? UNKNOWN

CONE CROP ASSESSMENT

Figure 14.2 Estimated population size of the Mt. Graham red squirrel for spring and autumn from 1986 to 1994 and estimates of cone crop data for those years. VATT is the Vatican Advanced Technology Telescope, SMT is the Submillimeter Telescope, and LBT is the Large Binocular Telescope.

MGIO complex has had excessive impact on the red squirrel population rests with this agency.

General Population Trends

During the spring and autumn of each year, biologists of the Forest Service, the Arizona Game and Fish Department, and the University of Arizona conduct a census. They visit a random sample, stratified by habitat, of about half of all the known midden sites on the mountain range to determine which are occupied. Investigators use this sample to estimate the size of the entire population.

When monitoring began in 1989, the red squirrel population in the Pinaleños had declined considerably since the first census conducted in spring 1986 (Coronado 1988). From a high of 348 ± 55 in spring 1986 the estimated population size had dropped to 235 ± 40 in autumn 1989 and was at its lowest recently recorded number of 161 ± 24 in the spring 1990 census. Coinciding with a bumper (mast) crop of Engelmann spruce, Douglas-fir, corkbark fir, and white fir cones in 1990, the population recovered to an estimated 263 ± 09 individuals by autumn 1990. The population continued to increase through 1991, reaching its highest recently recorded size of 390 ± 23 in autumn 1991. Since then, the population has declined and increased in concert with the available cone crops. In autumn 1993, following another good cone crop, the squirrel population was again larger than in 1986 (375 ± 32) and showed virtually no decline overwinter to spring 1994 (figure 14.2).

Red Squirrel Populations on the Monitored Areas

The red squirrel populations on the monitored areas reflect the changes seen in the entire population. These populations were at their lowest number in June 1990 with only 23 adults and 3 juveniles being found on approximately 327 ha. Coinciding with the good cone crop of 1990, the monitored populations increased to 89 adults and 3 juveniles by the end of that year and, after declining to 80 adults in June 1991, increased to 124 in November. The populations declined throughout 1992 but recovered to 114 squirrels by December 1993 (figure 14.3). In June 1994 the populations stood at 90 adults and 7 juveniles.

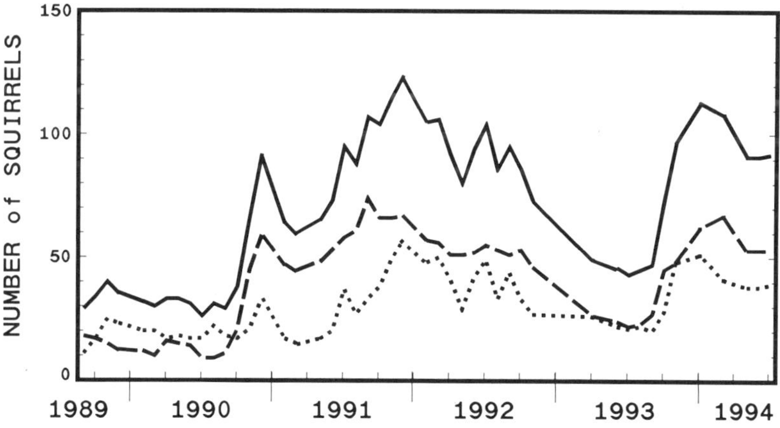

Figure 14.3 Population sizes of the red squirrel on the study areas showing the total population (solid line) and the populations within montane conifer (dashed line) and subalpine fir (spruce-fir; dotted line) habitats.

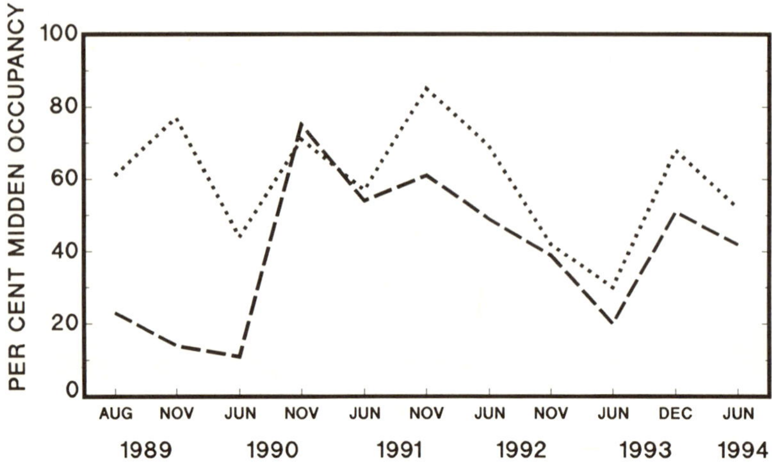

Figure 14.4 Spring and fall population densities for red squirrels on the study areas by habitat types. The dashed line is for montane conifer habitat populations, and the dotted line is for subalpine fir populations.

The crude density of red squirrels (number of squirrels per hectare) on the monitored areas is low. When the monitored populations were at their peak in 1991 the density was 0.38 squirrel/ha. This density is more in line with densities expected in the poorest red squirrel habitats (hardwood and mixed coniferous-hardwood forests) than in the spruce-fir and mixed-conifer forests on the Pinaleños. Spruce forests are generally considered to be the best red squirrel habitat (Obbard 1987). However, when densities are compared between habitats on the Pinaleños the spruce-fir habitat has had consistently lower densities of red squirrels than the upper-elevation mixed-conifer habitat, with a single brief exception in 1990 (figure 14.4).

Comparisons Between Construction and Nonconstruction Populations

Midden Occupancy

The only construction activity that occurred in 1989 involved the clearing and grading of the access road from Swift Trail (state highway 366) to the telescope sites in October and November. At the time only five red squirrel middens lay within 100 m of the access road and telescope sites. None of these was occupied before construction started, but one became occupied by the time construction ended for the year. The proportion of middens occupied on the construction areas was not significantly different from that on the nonconstruction areas before construction began or after it ended in 1989, nor has it differed in spring or autumn of any year since then, except in June 1994 in spruce-fir habitat (table 14.1) when the proportion of middens occupied on the SFC area was significantly greater than on the SFN (nonconstruction) area. The proportion of middens occupied in mixed-conifer habitat has typically been greater than that in the spruce-fir habitat (table 14.2).

Analyzing the possible effects of construction on the occupancy status of middens relative to their distance from construction is complicated by the small number of middens that are close to construction. The access road and telescope sites were deliberately placed as far away from known middens as possible. In addition to the midden that became occupied during construction in 1989, six new middens have been established within 100 m of construction since then. The average distance from construction of all middens on the construction areas is about 200 m (table 14.3), and the distance from construction for occupied versus unoccupied middens has not differed significantly in any year since 1989, except for 1992 (table 14.4). In autumn 1992 when the monitored squirrel populations were declining, middens that were closer to the road and telescope sites were more likely to be unoccupied than those farther away. We have not yet determined whether this pattern represents an effect of construction activity, was due to other factors such as habitat quality, or was simply the result of random loss of squirrels from the population. An analysis of key habitat structure components (canopy cover, foliage volume, downed log volume, and density of snags) for 27 of the middens on the construction areas showed that snag density increased as distance from construction increased (Young et al. 1994). Snags (standing dead trees) are preferred for nesting sites by red squirrels, and the increasing density of snags with distance from construction may reflect a general difference in the quality of the habitat as well. We did not include all of the middens on the construction areas in the analysis on snag density and are conducting further habitat work to determine whether the quality of the habitat may have a major influence on the occupancy of those middens closest to construction.

Additionally, the midden sites within 100 m of construction tend to be more isolated and have fewer other middens around them (mean local density of middens = 1.7) than those farther from construction (mean local density of middens = 3.0). In past years midden sites with

Table 14.1 Midden Sites on Each of the Monitored Areas Since 1989

	MCC		Within MC	MCN		SFC		Within SF	SFN	
	# Middens	% Occupied	*P*	# Middens	% Occupied	# Middens	% Occupied	*P*	# Middens	% Occupied
Aug. 89	12	92	—	—	—	26	23	0.977	53	23
Nov. 89	22	73	0.416	8	88	27	22	0.268	51	12
Jun. 90	22	45	0.305	10	40	27	10	0.972	52	10
Nov. 90	25	64	0.737	17	82	31	55	0.317	63	67
Jun. 91	26	58	0.902	18	56	34	56	0.802	68	53
Nov. 91	44	86	0.737	23	83	37	69	0.632	72	60
Jun. 92	44	70	0.674	23	65	37	51	0.678	72	39
Nov. 92	46	43	0.736	23	39	38	40	0.951	72	39
Jun. 93	46	33	0.595	23	26	38	13	0.203	72	24
Dec. 93	51	71	0.508	24	58	44	59	0.203	77	47
Jun. 94	50	52	1.000	25	52	57	51	0.016	78	28

Note: "*P*" values are for chi square tests on percent occupancy of construction versus nonconstruction areas within each habitat. Values less than 0.05 were accepted as statistically significant.

Table 14.2 Number and Occupancy Rates of Middens in Different Habitat Types on the Monitored Areas

	Mixed-Conifer Habitat			Spruce-Fir Habitat	
	# Middens	% Occupied	*P*	# Middens	% Occupied
Aug. 89	12	92	<0.000	79	23
Nov. 89	30	77	<0.000	78	14
Jun. 90	32	44	0.001	79	11
Nov. 90	42	71	0.242	94	63
Jun. 91	44	57	0.755	102	54
Nov. 91	67	85	0.001	109	61
Jun. 92	67	69	0.010	109	43
Nov. 92	69	42	0.705	110	39
Jun. 93	69	30	0.125	110	20
Dec. 93	75	68	0.023	121	51
Jun. 94	75	52	0.034	135	42

Note: "*P*" values are for chi square tests on percent occupancy of mixed-conifer versus spruce-fir habitat. Values less than 0.05 were accepted as statistically significant.

greater local densities of middens have tended to be more continuously occupied than middens with low local densities (Young et al. 1993).

Density of Squirrels

The local density of squirrels on the construction areas has been either similar to or greater than that on the nonconstruction areas at the end of each year since 1989 (table 14.5). Middens in mixed-conifer forest typically have three to four other middens within 100 m (e.g., 3.67 ± 0.23, N = 75, Dec. 1993), while middens in spruce-fir forest rarely have more than two other middens within 100 m (e.g., 2.2 ± 0.14, N = 121, Dec. 1993). Since 1989, the local density of squirrels in mixed-conifer forest has ranged from one to four, whereas in spruce-fir forest it rarely exceeds one (Young et al. 1994).

Female Reproductive Success

We have not been able to obtain precise estimates of female reproductive success. Biologists observed no litters on the monitored areas from August through November 1989, but two litters of unknown size were seen on the MCC area and one litter of unknown size was seen on the SFC

Table 14.3 Mean Distance from Construction-Related Disturbances of All Middens on the Construction Areas, 1989–1993

	November 1989			November 1990			November 1991			November 1992			December 1993		
Habitat	N	$\overline{X}$	s.e.	N	$\overline{X}$	s.e.	N	$\overline{X}$	s.e.	N	$\overline{X}$	s.e.	N	$\overline{X}$	s.e.
MCC	22	220.1	14.8	25	218.3	13.9	44	208.6	11.4	46	209.2	11.0	51	212.1	10.7
SFC	27	217.6	22.4	31	216.3	19.9	37	210.2	18.3	38	214.9	18.4	44	205.9	16.6

Note: Some middens are more than 300 m from construction but were included in the construction areas due to errors in locations on the original midden maps. N = number of middens; $\overline{X}$ = mean distance (meters); s.e. = standard error.

Table 14.4 Mean Distance (meters) from Construction-Related Disturbances of Occupied and Unoccupied Middens on the Construction Areas, 1989–1993

Habitat	Midden Status	November 1989			November 1990			November 1991			November 1992			December 1993		
		N	$\overline{X}$	*P*	*N*	$\overline{X}$	*P*	*N*	$\overline{X}$	*P*	*N*	$\overline{X}$	*P*	*N*	$\overline{X}$	*P*
MCC	unoccupied	7	200.1 ±27.4		9	213.9 ±22.2		6	177.3 ±40.6		27	189.2 ±14.1		15	204.1 ±21.1	
				0.368			0.819			0.281			0.028			0.638
	occupied	15	229.5 ±17.6		16	220.7 ±18.2		38	213.5 ±11.6		19	237.7 ±15.7		36	215.4 ±12.6	
SFC	unoccupied	22	224.3 ±24.9		14	193.1 ±28.6		13	197.4 ±31.0		23	186.5 ±24.1		18	194.7 ±27.9	
				0.543			0.298			0.616			0.055			0.581
	occupied	5	188.3 ±54.8		17	235.4 ±27.6		24	217.0 ±23.0		15	258.5 ±25.4		26	213.7 ±20.7	

Note: "*P*" values are for analysis of variance on distance from construction of occupied versus unoccupied middens. Values less than 0.05 were considered to be statistically significant. *N* = number of middens; $\overline{X}$ = mean distance (meters).

Table 14.5 Mean local density (± 1 s.e.) of Red Squirrels on Each of the Monitored Areas, 1989–1993

	Mixed-Conifer Habitat		Spruce-Fir Habitat	
	MCC	MCN	SFC	SFN
Nov. 89	1.5 ± 0.34 a	1.7 ± 0.36 a	0.0 ± 0.0 b	0.3 ± 0.20 b
Nov. 90	2.2 ± 0.40 a	2.9 ± 0.39 a	1.3 ± 0.19 b	1.1 ± 0.17 b
Nov. 91	3.5 ± 0.33 a	2.5 ± 0.31 b	1.5 ± 0.18 c	1.1 ± 0.14 c
Nov. 92	1.3 ± 0.18 a	1.3 ± 0.47 a	1.2 ± 0.28 a	0.6 ± 0.10 a
Dec. 93	3.4 ± 0.27 a	2.4 ± 0.36 b	1.8 ± 0.22 b	1.1 ± 0.17 c

Note: Differences among areas were tested for statistical significance by analysis of variance and SNK multiple-range tests; numbers with the same letter following them in each row are not statistically different.

area earlier in the summer before monitoring began. Moreover, we did not make accurate counts of the total number of females on each monitored area in 1990; we estimate that for most of 1990 the SFN area had no females and the SFC area had only one. We observed six litters on the combined mixed-conifer forest areas (mean litter size 2.67) and one litter of unknown size on the SFC area.

Since 1991 we have made a concerted effort to locate all of the females on each area and to determine their reproductive condition regularly. When we observe a lactating female, we visit her every few days in an attempt to determine where her natal nest is located and to count the number of juveniles as soon after their emergence as possible. Not all females that lactate are successful at rearing their litters to emergence; however, the number of litters observed is no doubt an underestimate of reproductive success, as we do not locate some natal nests and do not see the litters.

The proportion of lactating females varies greatly from one area to another and from year to year. Overall the proportion of females observed to be lactating each year has ranged from 18% to 89% (table 14.6). No litters were observed on the SFN area in 1990, but otherwise the proportion of females rearing litters has ranged from 10% to 42% on all areas (table 14.6). The proportion of females lactating, and those rearing litters, has not differed significantly between construction and nonconstruction areas in any year since 1990 (Young 1991, 1992; Young et al. 1993, 1994). Large variation in reproductive effort and reproductive success is common in red squirrel populations, due to variation in the proportion of females that breed and variation in litter size from year to year (Obbard 1987).

Litter sizes have also varied between areas and from year to year (table 14.6). The number of litters found each year is small, making comparisons between areas difficult, but it appears that there is no consistent difference in the average litter size on construction versus nonconstruction areas or between populations in different habitats. The average litter size for all years and areas combined is 2.13 (range 1–5, N = 45) somewhat smaller than reported for other red squirrel populations (Obbard 1987). In some years, a few females appear to have produced two litters per year. Although multiple annual litters are

Table 14.6 Estimates of Female Reproductive Success on the Monitored Areas, 1990 through 1993

		MCC	MCN	SFC	SFN	Total
1990						
	# ♀	—	—	—	—	—
	% lactating	—	—	—	—	—
	% observed w/litters	—	—	—	—	—
	mean litter size (N)	2.67 (3)	2.67 (3)	—	—	2.67 (6)
1991						
	# ♀	10	7	12	30	59
	% lactating	80	71	67	50	63
	% observed w/litters	30	29	42	17	25
	mean litter size (N)	2.67 (3)	2.00 (2)	1.80 (5)	1.60 (5)	2.33 (15)
1992						
	# ♀	28	10	17	27	82
	% lactating	32	40	18	30	29
	% observed w/litters	11	33	12	18	16
	mean litter size (N)	2.33 (3)	1.67 (3)	3.00 (2)	2.00 (5)	2.15 (13)
1993						
	# ♀	16	9	9	10	44
	% lactating	50	33	89	40	50
	% observed w/litters	31	22	33	10	25
	mean litter size (N)	1.60 (5)	2.00 (2)	2.67 (3)	3.00 (1)	2.09 (11)

Note: No litters were observed from August through November in 1989, and the total number of females was unknown in 1990. Numbers in parentheses are the number of litters observed.

Table 14.7 Reproductive Success of Females on the Monitored Construction Areas in 1993, Relative to Distance from Construction

Breeding Status	*N*	Distance from Construction (meters) Mean	Minimum	Maximum
Not breeding	9	227.1	122.5	352.8
Lactated/no litter emerged	8	155.2	45.6	309.9
Lactated/litter emerged	8	226.2	39.6	385.7
Kruskal-Wallis test : $p = 0.158$				

uncommon in most red squirrel populations, red squirrels in the southeastern portion of their geographic range have also been reported to produce two litters per year (Klugh 1927; Hamilton 1939; Layne 1954; Lair 1985), and one previous report documents females having two litters in one year on the Pinaleños (Froehlich 1990).

In 1993, after accurately measuring the distance of each midden from the nearest construction disturbance, we analyzed the reproductive success of females on the MCC and SFC areas relative to their distance from construction. The relationship between reproductive success and distance from construction was not statistically significant (table 14.7); females as close as 40 m to construction areas bred, and in 1993, a female occupying the closest midden to construction reared her litter to emergence.

Implications of Change in Red Squirrel Populations

All of the changes in red squirrel populations on the Pinaleño Mountains since 1989 can so far be attributed to variation in the abundance and distribution of food resources. Although biologists have collected no quantifiable data on food resources, the Forest Service has made a qualitative assessment of the conifer seed crops since 1989 (figure 14.2). Bumper crops of Engelmann spruce have occurred twice since then, in 1990 and 1993. Cone crops of other conifers (notably, Douglas-fir, corkbark fir, and Western white pine) have been good to excellent in some areas during the intervening years. The red squirrel population responded to the high cone crops of 1990 and 1993 with large increases in number. The monitored population also increased in other years but on a local level that did not coincide with large increases in the total population. The only year-to-year drop in the total population occurred in 1992, when the cone crops of all species except white fir were not good (figure 14.2).

To date, no conclusive evidence demonstrates that the construction of the MGIO complex on the Pinaleño Mountains has had any adverse affect on the nearby red squirrel population. The squirrel populations on the construction areas have increased and decreased in concert with populations on nonconstruction areas; reproductive success and litter sizes are virtually identical on construction and nonconstruction areas; and in some respects, such as density of squirrels, the populations on construction areas appear to be in better condition.

One of the greatest problems confronting the monitoring effort is the small number of red squirrels living in proximity (< 100 m) to the observatory site (and the small population size in the Pinaleños generally). Because of our small sample sizes, we have not been able to provide a definitive explanation for the one aspect of the red squirrel populations on the construction areas that suggests a potential impact of construction activity, the apparently decreased probability that a midden close to construction will be occupied. However, the establishment of new middens near construction (six within 100 m since 1989) and the continued occupancy of several of these middens through the intense construction in 1990 through 1992 (figure 14.2) suggests that construction activity is not a major factor influencing the use of nearby middens.

To analyze patterns of population changes more accurately in the future, especially for specific local areas, we have undertaken several long-term monitoring projects to provide data on the suitability of the forest habitats for red squirrels living on the monitored areas. In 1993, we began collecting data on conifer seed crops and epigeous (above-ground) mushroom crops from 34 randomly distributed plots on the monitored areas (Young et al. 1994). These data will eventually be useful for interpreting changes in squirrel numbers and distribution relative to available food resources and will enable us to more easily separate the effects of construction from other environmental effects.

In 1993, we also began collecting data on habitat attributes (plant species, physical characteristics) around newly established middens. These data, combined with data on the occupancy of middens, provide the basis for an analysis of the habitat preferences of red squirrels and an assessment of the quality of individual midden sites. The habitat structure data will eventually be expanded to include a detailed assessment on all of the monitored areas and will thus enable us to determine the relative quality of each as red squirrel habitat.

In 1994, we established two meteorological stations on the monitored areas to provide data on local year-round weather conditions.

During the summers of 1990 through 1992, we also conducted a small study to examine the effects of the new access road to the MGIO site on the microclimate of the forest floor adjacent to the road. Forest Service biologists predicted that edge effects from new clearings in the forest would result in the loss or degradation of an area of red squirrel habitat up to four times greater than the actual area cleared (Coronado 1988), but no studies were originally planned to verify this prediction. A pilot study in 1990 along the entire access road suggested that edge effects on soil temperature and soil moisture of the forest floor extended less than 25 m into the forest (Young, unpublished data); the study thereafter concentrated on a short section of the road in homogeneous spruce-fir habitat on the upper end of the road. We measured soil temperature and soil moisture weekly (at a depth of 15 cm) at 5 m intervals along ten 50 m transects extending perpendicularly to either side of the road during the summers of 1991 and 1992. A method for measuring an index of evaporation was added to the study in 1992. The preliminary analysis of the data from this study confirmed that the edge effects of the road extended less than 25 m into the forest. In fact, soil moisture appears to not be affected at all, and soil temperature was elevated by only a few degrees at the very edge, and for 5 m into the forest, and only on the uphill (north) side of the road.

The results of this study and a similar one conducted around the telescope sites suggest that the loss of red squirrel habitat due to the construction of the MGIO complex has been far less than anticipated in the expanded biological assessment (Coronado 1988). However, these studies are far from an exhaustive examination of edge effects, and further study of this question is needed (Kreisel and Young 1994).

Of more limited scope, but of no less importance than monitoring for construction impacts, is a continuing effort by monitoring biologists to conduct research on the comparative ecology of the Mt. Graham red squirrel and other populations of red squirrels in the southwestern United States. These efforts are limited because most of our time and personnel must be devoted to the close scrutiny of the populations on the monitored areas. Nevertheless, 1994 saw the completion of a small study that examined differences in the use of different nest types, particularly grass nests, between populations of red squirrels in mixed-conifer and spruce-fir forests on the Pinaleño and White Mountains of Arizona (Six and Young 1994). Future comparative research involving Pinaleño and White Mountains populations is planned. Biologists conducting these studies will compare habitat structures and the abundance, distribution, and use of food resources. Other southwestern populations will also be studied if possible. The results of this program of monitoring and research should eventually help us understand the causes of the endangerment of the Pinaleños red squirrel population and provide a sound basis for its future protection.

References

Chase, C. P., and D. E. Brown. 1982a. Rocky mountain (Petran) subalpine conifer forest. In *Biotic Communities of the Southwest–United States and Mexico,* ed. D. E. Brown, pp. 37–39. University of Arizona Press, Tucson. Special issue of *Desert Plants* 4.

Chase, C. P., and D. E. Brown. 1982b. Rocky mountain (Petran) and Madrean montane conifer forests. In *Biotic Communities of the Southwest–United States and Mexico,* ed. D. E. Brown, pp. 43–48. University of Arizona Press, Tucson. Special issue of *Desert Plants* 4.

Coronado National Forest. 1988. *Mount Graham Red Squirrel: An Expanded Biological Assessment.* Coronado National Forest, Tucson, Ariz.

Coronado National Forest. 1989. *Management Plan for the Mount Graham International Observatory.* Coronado National Forest, Tucson, Ariz.

Czaran, T., and S. Bartha. 1992. Spatio-temporal dynamic models of plant populations and communities. *Trends in Ecology and Evolution* 7:38–42.

Froehlich, G. F. 1990. Habitat use and life history of the Mt. Graham red squirrel. M.S. thesis, University of Arizona, Tucson.

Hamilton, W. J., Jr. 1939. Observations on the life history of the red squirrel in New York. *American Midland Naturalist* 22:732–45.

Kemp, G. A., and L. B. Keith. 1970. Dynamics and regulation of red squirrel (*Tamiasciurus hudsonicus*) populations. *Ecology* 51:763–79.

Klugh, A. B. 1927. Ecology of the red squirrel. *Journal of Mammalogy* 35:252–53.

Kreisel, W. E., Jr., and P. J. Young. 1994. Edge effects along a new forest road. First Annual Wildlife Society Conference, Albuquerque, N.M. (Unpublished abstract.)

Lair, H. 1985. Mating seasons and fertility of red squirrels in southern Quebec. *Canadian Journal of Zoology* 63:2323–27.

Larsen, K. W., and S. Boutin. 1994. Movements, survival, and settlement of red squirrel (*Tamiasciurus hudsonicus*) offspring. *Ecology* 75:214–23.

Layne, J. N. 1954. The biology of the red squirrel, *Tamiasciurus hudsonicus loquax* (Bangs), in central New York. *Ecological Monographs* 24:227–67.

Obbard, M. E. 1987. Red squirrel. In *Wild Furbearer Management and Conservation,* ed. M. Novak. Ontario Trappers Association and Ontario Ministry of Natural Resources, Toronto.

Price, K., K. Broughton, S. Boutin, and A. R. E. Sinclair. 1986. Territory size and ownership in red squirrels: Response to removals. *Canadian Journal of Zoology* 64:1144–47.

Six, S. K., and P. J. Young. 1994. The occurrence of grass nests in Mount Graham red squirrel middens. First Annual Wildlife Society Conference, Albuquerque, N.M. (Unpublished abstract.)

Smith, A. A., and R. W. Mannan. 1994. Distinguishing characteristics of Mount Graham red squirrel midden sites. *Journal of Wildlife Management* 58:437–45.

Spear, M. J. 1988. *Biological Opinion.* U.S. Department of the Interior, Fish and Wildlife Service, Albuquerque, N.M.

Spicer, R. B., J. C. DeVos, and R. L. Glinski. 1985. Status of the Mt. Graham red squirrel, *Tamiasciurus hudsonicus grahamensis* (Allen), of southeastern Arizona. U.S. Department of the Interior, Fish and Wildlife Service, Albuquerque, N.M. (Unpublished.)

Vahle, J. R. 1978. Red squirrel use of southwestern mixed coniferous habitat. M.S. thesis, Arizona State University, Tempe, Ariz.

Young, P. J. 1991. Annual report for 1990 of the University of Arizona—Mt. Graham red squirrel monitoring program. (Unpublished.)

Young, P. J. 1992. Annual report for 1991 of the University of Arizona—Mt. Graham red squirrel monitoring program. (Unpublished.)

Young, P. J., M. I. Camp, V. L. Greer, W. E. Kreisel Jr., and M. A. Martin. 1993. Annual report for 1992 of the University of Arizona—Mt. Graham red squirrel monitoring program. (Unpublished.)

Young, P. J., M. I. Camp, V. L. Greer, W. E. Kreisel Jr., M. A. Martin, and S. K. Six. 1994. Annual report for 1993 of the University of Arizona—Mt. Graham red squirrel monitoring program. (Unpublished.)

CHAPTER FIFTEEN

Overview: Reflections on Prehistoric Turbulence

Paul S. Martin

What do those who care need to know about the historical ecology of Arizona's mountain (sky) islands, including Mt. Graham? As we have seen, the interested parties are many. Besides astronomers there are bible campers, biologists, bird watchers, Earth Firsters, hunters, personnel of the Fish and Wildlife Service, Forest Service, Arizona Game and Fish Department, and other governmental agencies; some are litigants in court or their allies. I have barely scratched the surface. Those concerned with the recent storm or "train wreck" in the Pinaleño Mountains, which generated this book, may also be curious about earlier turbulence. The squirrels, packrats, voles, and black bears, to name a few, that live in these mountains have endured as species for hundreds of thousands of years, their lineages for millions. The trees—spruce, corkbark fir, white fir, Douglas-fir, pines, oaks, and maples—evolved more slowly. Their pollen types and leaves or fruits can be detected in fossil deposits tens of millions of years old. What kinds of storms have the animals and plants on Mt. Graham and the isolated range it occupies (the Pinaleños) weathered in the past, in particular during the last 40,000 years, the time slice that can be dated by radiocarbon?

Before we probe deeper into the past, it is worthwhile to reflect on the more recent history of southern Arizona. The first colonists from the Old World were Paleoindians who came from Asia more than 10,000 years ago. In historic times the first European to traverse the region was the castaway Álvar Núñez Cabeza de Vaca. Pima Indians joined his entourage on its way into Sinaloa, where Cabeza de Vaca's accounts eventually prompted the expedition to New Mexico (ultimately to Kansas) of Francisco Vásquez de Coronado. The first "Spaniard" to appear in southern Arizona on the San Pedro River, within 160 km of Mt. Graham, was Estevan, who was guiding and in advance of Fray Marcos de Niza. Estevan, slave of Andrés Dorantes in Cabeza de Vaca's party, was a Moor rather than a Spaniard.

On the San Pedro the Sobaipuri Pimas irrigated maize, beans, and cotton. In 1697 and 1698 they defeated the Apaches, Jocomes, and Janos in pitched battles (Bolton 1936). Nevertheless, around the Pinaleños and adjacent mountain islands the Apaches had the upper hand, and the Sobaipuri had to withdraw to merge with other Pimas to the west (Spicer 1962).

In the nineteenth century those Anglos who traveled through or (less often) settled in the borderlands of southern Arizona, southern New Mexico, and west Texas did not come for the mountains or the biotic and cultural diversity. In 1853, following the war with Mexico, the Gadsden Purchase (locally the strip from 31°20′N latitude and northward to the Gila River) added two dozen sky islands, many the home of Apaches, to southern Arizona. With no thought of astronomy, biogeography, or emerald cienegas, the United States wanted the strip so that the nation could secure an all-season transcontinental railroad corridor for the southern states.

By the time railroad construction reached Arizona 30 years later, the Apaches had been concentrated on the San Carlos Reservation, north of the Pinaleños, minus irrigated lands appropriated by white farmers along the Gila River. Developers of new copper mines operated without regard to the boundaries of the reservation (Spicer 1962). The Pinaleños were no longer Apache territory.

With the completion of the railroad the cattle industry boomed, only to go bust in the drought of the 1890s. Natural sky island parklands and cienegas were grazed and lumbering began. Recently geographer Conrad Bahre discovered a 1902 map of the Chiricahua Mountains drawn by Albert E. Potter at a scale of 1½ miles to the inch. The map locates 11 sawmills and extensive areas of cut timber between 6,000 and 9,000 feet (1,800–2,700 m). At the time the devastation on the larger sky islands, including Mt. Graham, resembled what one may find now around old lumber camps in the Sierra Madre Occidental of Chihuahua, Mexico (Bahre, personal communication 1994; Bahre 1991:167). Perhaps that is why no early naturalists such as Mary Austin, John Muir, or Ernest Thompson Seton who revealed the beauty of other parts of the West wrote about the ecological treasures of the Gadsden Purchase or the glories of its mountain islands. The place was a wreck.

In 1903, with the establishment of the Desert Laboratory outside Tucson, botanists of the Carnegie Institution of Washington began to analyze plant distributions on desert mountains, especially the Santa Catalinas (Shreve 1915). In 1914 Forrest Shreve of the Desert Laboratory and John James Thornber made the first extensive plant collections on Mt. Graham (McLaughlin 1993a), noting the unique character of its high-elevation flora (Shreve 1919). An inadvertent outcome of the Gadsden Purchase was the division of sky islands into different management units. Those mountains north of the U.S.-Mexico border came under fire control and are now at risk of fire storms as a result of excess deadwood, whereas those south of the border were left alone to burn naturally and often. The different regimes have had a profound effect on the structure of pine-oak woodlands and their native breeding birds (Marshall 1957). Severe fuel buildup in the forests surrounding the observatory poses a serious management problem for both astronomers and conservationists.

A quantitative account of the vegetation gradient on the Santa Catalinas (Whittaker and Niering 1965, 1975), widely reprinted in many ecological textbooks, incorporated data from the top of the Pinaleños, that is, the spruce-fir forest above 2,700 m. The result is a detailed analysis of plant distribution on a hypothetical desert mountain from the bottom to the top, from saguaros to spruce (Whittaker and Niering 1964). The Pinaleños harbor some 900 species of vascular plants (McLaughlin 1993b), the Chiricahuas, over 1,200. Other researchers have investigated the biogeographic and ecological attributes of the borderland mountains (Marshall 1957; Gehlbach 1993) and recognized the importance of post-Pleistocene immigrations (Lomolino et al. 1989) and extinctions (Gehlbach 1993). Nevertheless, astronomers put Mt. Graham on the map and into the national headlines.

Millions of people stream through southern Arizona just south of Mt. Graham and the Pinaleños. On most days around the town of Willcox passengers in vehicles on Interstate-10 (I-10) are in clear view of not only the Pinaleños, but also the Chiricahuas, Dos Cabezas, Santa Catalinas and Rincons, five of the eight highest desert mountains in southern Arizona (table 15.1). Nevertheless, the montane forests of interest in this book cannot be seen from the interstate. I suspect that, even if told, most travelers would not believe that dense, mossy, old-growth spruce and fir trees exist on mountains within 50 to 100 km of their route. Nor would their eyes tell them that the area from southeastern Arizona into southwestern New Mexico is considered a biogeographic melting pot, richer in species of small mammals, reptiles,

Table 15.1 Highest Island Mountains of the Gadsden Region, Southern Arizona

Mountains	Highest Peak	Elevation (E) (feet)	Base (B) (feet)[b]	Relative Relief (E-B) (feet)[b]	Relative Relief (E-B) (meters)[b]	Area Above 8,000 Feet (square miles)	Higher Plant Taxa
Pinaleños[a]	Mt. Graham	10,720	4,000	6,700	2,000	32.0[c]	834
Santa Catalinas[a]	Mt. Lemmon	9,157	3,000	6,100	1,900	5.9	843
Rincons	Mica Mountain	8,666	3,000	5,700	1,700	—	864
Santa Ritas[a]	Mt. Wrightson	9,453	4,000	5,400	1,700	1.8	—
Chiricahuas	Chiricahua Peak	9,796	4,500	5,300	1,600	26.9	1,199
Huachucas	Miller Peak	9,466	5,000	4,500	1,400	7.2	—
Dos Cabezas	Dos Cabezas	8,369	4,000	4,400	1,300	—	—
Baboquivaris[a]	Baboquivari Peak	7,730	3,500	4,200	1,300	—	—

Source: After McLaughlin 1993b and Warshall 1986; data on higher plant taxa after Bennett and Kunzmann 1992. "Higher plants" are trees, shrubs, forbs, grasses, and ferns.

[a] Astronomical site.

[b] Data rounded.

[c] 2 mi^2 above 10,000 feet.

ants, butterflies, and many other kinds of organisms than are found elsewhere in the United States or Canada.

The number of travelers is large. In 1990, 1991, and 1992 the vehicle count along I-10 in eastern Arizona, from Texas Canyon to San Simon near the New Mexico line (minus local traffic) varied from 8,000 to 10,000 vehicles daily, according to Arizona Department of Transportation data. That count multiplies to over 3 million passenger cars, trucks, and buses a year, not to mention AMTRAK passengers on the Southern Pacific and air traffic into and out of borderlands cities. If anyone bothers to stop, it is seldom to admire the Pinaleños, the Chiricahuas, and the other flagships of the sky island fleet, much less the biotic diversity they harbor.

Around Willcox the transportation corridor looks bleak. Here one finds bunchgrass (alkali sacaton), saltbush, inkweed, and jackass clover, punctuated with low, frost-pruned patches of Torrey mesquite. The season of green grass is limited to three months of "monsoon" rains from mid July to mid October. Summer storms can turn the Willcox Playa into a shallow lake supporting a bloom of floating algae devoured by emergent crustaceans such as fairy shrimp, clam shrimp, and tadpole shrimp. Before May, the winters, no matter how wet, will not induce appreciable new growth in sacaton or mesquite or crustaceans. Much of the year the color of the landscape is a dirty blond, and if the playa looks wet it is a mirage. Willcox, like other localities in playa basins or interior valleys, lies in a cold-air sump. On clear, still nights cooling air settles into the middle of the valley below 1,300 m. The mean monthly low temperature at Willcox averages 8°F (6°C) *less* than the monthly lows for adjacent stations that are on slopes tens to hundreds of meters higher in elevation (Fort Grant, Tombstone, Bisbee). The record low at Willcox of −22°C (−7°F) occurred 7 December 1978, the day after an arctic air mass brought enough snow to close I-10 between Tucson and the New Mexico border (Sellers et al. 1985). As a result, Torrey mesquites were killed to the ground.

Barren as it is, the Willcox Playa is the ideal spot from which to start reflecting on deep history and the changing destiny of the sky islands. One strong hint of ancient transformation is visible in the shoreline itself, a curving berm, or beach ridge, that delivered sand east of Willcox when the playa was a lake. An early artifact found here, a Clovis point, indicates human presence over 10,000 years ago (Waters and Woosley 1990). The beach sands come mainly from the Pinaleños, the Chiricahuas, and to a lesser degree the Winchesters, a much smaller, seldom-visited mountain island to the northwest of Willcox that is barely big enough to harbor its own pines, Douglas-fir, and hard maple, although no one driving the interstate would suspect this.

The native trees seen easily from vehicles on I-10 appear at exit 322 west of Willcox, at an elevation of 1,500 m (5,000 feet), the highest point on the route. Here, as one enters Texas Canyon, is an encinal (Spanish for "oak woodland"), a rich mix of evergreen round-crowned Emory oaks, sotol, beargrass, century plants, and many other shrubs scattered between scenic granite domes. This landscape is often seen in old movies; we may recognize it as the spot where the stagecoach is robbed.

As Apaches know, the acorns of Emory oaks are edible, and when trimmed and roasted the hearts of century plants and sotol yield sugary "cabezas" or heads. Although encinal does not extend much farther north, a great deal of it occupies the skirts of the mountain islands of Cochise County, Arizona; southward through northeastern Sonora; and, on the other side of the continental divide, down the dry (eastern) side of the Sierra Madre Occidental of Chihuahua, Durango, Zacatecas, and Jalisco, Mexico. Portions of all the larger sky islands in southern Arizona support a belt of encinal. In the encinal and above in pine-oak woodland in the Pinaleños are 10 species of oaks and 5 of pines (McLaughlin 1993a). Numbers of species of oaks and pines increase in the Sierra Madre. For example, in pine-oak habitat about 500 km to the south in western Chihuahua, one finds 20 species of oaks. Within a few miles of the 240 m high waterfall at Basaseachic on the headwaters of the Rio Mayo, Richard Spellenberg has found 10 species of pines (personal communication 1994). The distinctive Mexican pine-oak biota, whether comprising oaks, pines, songbirds, or fence lizards, to name a few ingredients, is much richer south of the border (Marshall 1957). Nevertheless, thanks to the sky islands of the Gadsden Purchase some fine examples of both encinal and pine-oak woodland along with a large chunk of the Madrean archipelago ended up in the United States.

Around Willcox, traffic down the interstate rides on road metal (fill) excavated from cross-bedded sand and gravel deposits of the ancient beaches. Occasionally the quarries yield unmistakable evidence of biotic change from the last ice age, especially bones of fossil mammoth, extinct camel, and extinct horse (Lindsay and Tessman 1974). North of Willcox, gravel-pit operators have shown me buckets of mammoth bones discovered during construction of the interstate highway, and I suspect many molars of extinct Columbian mammoth were unintentionally ground up with the aggregate to make the roadbed of I-10. Travelers might also be bemused to learn that between exits 340 and 331 west of Willcox, their route lies below the shoreline of Pluvial Lake Cochise. The lake was 12 m deep (Waters 1989); Willcox, I-10, and the railroad would have been under several meters of water as late as 13,000–14,000 years ago, had these human constructs existed then.

In the barren playa, a few feet beneath tight impermeable clays, hydrologist Oscar Meinzer discovered a deep deposit of black mud, mainly illite—the Quaternary sediments of Lake Cochise (Meinzer and Kelton 1913). To an eastern limnologist, black mud suggests a high organic content, which is not the case here; the organic matter content of this mud is <2%. The mud smells of rotten eggs and is highly reduced (accounting for the color) with an oxidation-reduction potential of −500. It is unusually alkaline, having a pH around 9.0 (Martin 1965). This chemical environment is favorable for preserving fossil pollen and spores, remains of ostracods (small crustaceans), and colonies of algae of the genus *Botryococcus*.

A 42 m core into the lake muds yielded interesting variations in fossil pollen, suggesting that during various cold stages of the most recent lengthy glacial interval the Sulphur Springs Valley once supported woodlands—at times oak, juniper, and grasses, at times pine and sagebrush—in place of modern mesquite-grassland and its modern pollen rain of ragweed, saltbush, and grass. Near the top of the core the pollen of oak, juniper, grasses, and sagebrush gave way to 98% pine and 1–2% spruce and fir (Martin 1965), which likely reflected the existence of some sort of pine forest with patches of spruce near the shores of Pluvial Lake Cochise (Martin 1963). This is the interval most likely to have seen Mt. Graham red squirrels and spruce trees in contact with other populations, if contact occurred in the last 40,000 years.

According to radiocarbon dates on fossils, spruce occupied the top of the Pinaleños for at least the last 8,000 years (Anderson and Shafer 1991; Anderson and Smith 1993). Fossil pollen profiles show that a few thousand years earlier in central Arizona and New Mexico, spruce prevailed at 2,100 m, as at Potato Lake (Anderson 1993) south of Flagstaff. Potato Lake is now surrounded by ponderosa pine. During the late Pleistocene, spruce and red squirrels both reached the Guadalupe Mountains of western Texas, hundreds of kilometers from mountains where either now occur (Chapter 10). During the last cold stage around 18,000 years ago, when the spruce-fir community expanded, it may have been in a drier phase as an open woodland of scattered spruce and limber pine embedded in sagebrush. Whatever the structure of the community, spruce spread over parts of Arizona and New Mexico well below its modern elevations and beyond its historic range (Betancourt et al. 1990). During most of the last 400,000 years such cold conditions at times were probably the norm, not the exception (Porter 1989).

Farther to the south in the mountains of northern Mexico is a different species of spruce. Chihuahua spruce features a larger cone and a different needle chemistry than either blue spruce or Engelmann spruce, the species that are known north of the border and into the Rocky Mountains. Chihuahua spruce is relatively local in western Chihuahua and Durango, Mexico, occurring at 2,500–2,700 m and mixing with montane pines, oaks, fir, Douglas-fir, and occasional aspen; the community is Chihuahua's version of Arizona's mixed-conifer forest. Chihuahua spruce is not common; Mexican foresters have censused only some 16,000 individuals (Robert Bye, personal communication). The habitat looks suitable for red squirrels.

Presumably during the last cold stage, and earlier, Engelmann spruce occupied parts of northern Mexico. It survives there at only two localities, both above 3,600 m (Taylor et al. 1994). One is on the Sierra Mohinora, the highest mountain in the state of Chihuahua; the other is located 800 km to the east in the Sierra Madre Oriental outside Monterrey, Nuevo Leon, an area of extraordinary endemism (McDonald 1993). Neither stand is well known. Quite possibly they and other mixed-conifer stands in northern Mexico once supported red squirrels. North of the border, as Donald Hoffmeister concluded (1986:217), "some time in the not too distant past red squirrels must have been present on other mountain ranges in southeastern Arizona — Chiricahuas, Santa Catalinas, Huachucas — and they may have been most closely related to those in the Grahams."

Perhaps some of this history will help astronomers to understand why previous telescope construction on other mountain islands — Kitt Peak in the Baboquivari Mountains, Mt. Bigelow in the Santa Catalinas, and Mt. Hopkins in the Santa Ritas (none of the sites jeopardizing highly restricted mountaintop habitat or threatening endemic species) — occasioned scant outcry among environmentalists, whereas Mt. Graham in the Pinaleños was another story (see Warshall 1986). Even in the absence of endemic red squirrels, it would have been a serious matter to intrude on this isolated spruce–alpine fir forest, a forest whose interesting history is just coming to light.

Another significant feature of the region deserves attention. Southwestern Colorado, Arizona, and New Mexico are very rich in some of the most impressive archaeological features, including pueblos, ball courts, and irrigation systems, that exist in the United States (Cordell 1984), a spillover from the prehistoric treasures to be found in Mexico. A major museum, the Amerind Foundation, located west of Willcox at Dragoon, Arizona, exhibits some of the prehistoric legacy, and archaeologists associated with the museum conduct research on the region. Archaeologists have excavated pit houses, pottery, and grinding stones along midvalley floodplains and around playas. They find perishable remains (basketry, sandals, corncobs) in caves or rockshelters. Although archaeologists have not surveyed the Pinaleños in any detail, the top of Mt. Graham yielded rock cairns and a turquoise bead when potential telescope sites were inspected (John Madsen, personal communication 1994). The Mt. Graham artifacts, including Mogollon pottery, are not directly traceable to any extant organized tribe, although the Zuni may be lineal descendants of the Mogollon culture (Spoerl 1992). Still older Archaic sites from the Cochise Culture of 2,000–8,000 years ago are widely distributed among the mountain islands in Cochise County, including surfaces around Willcox Playa (Woosley and Waters 1990). One Archaic site on a 2,600 m peak in the Pinaleños yielded a collection of stemless points (Lee 1986).

Some of the oldest well-established archaeological records in the New World are found in buried sites along the San Pedro River between Fairbank and Naco, Arizona. The sites lie within 160 km of Mt. Graham. Here, around 11,000 years ago, as determined by repeated radiocarbon dating, early Americans using Clovis points and other stone and bone tools hunted and processed mammoth and bison (C. V. Haynes 1991, 1992). Clovis sites have been found in a small and intriguing number of localities between Wyoming and the Mexican border. Clovis foragers were the only people definitely known to encounter New World elephants, both mammoths and mastodons, both on the verge of extinction.

Archaeologists, geologists, and biologists alike have a good deal of interest in, and uncertainty about, whether Clovis hunters might have exterminated the mammoth (G. Haynes 1991). Many other large animals—including ground sloths, horses, camels, mastodons, tapir, extinct goats, peccary, and species of bison unlike living bison—also disappeared around the same time, although only mammoth, mastodon, and bison are in a convincing association with Clovis artifacts. A major controversy has arisen about whether climatic change, prehistoric colonizers, or both had a major role in the catastrophic extinction of several dozen genera of New World megafauna (Martin and Klein 1984). At first the controversy may seem moot to stewards of the Pinaleños since the crisis, whatever its nature, is past and the sky islands are no longer in the seasonal range of mammoth, tapir, and ground sloths. Nevertheless, extinctions past may well bear on those yet to come, and if the mountain cienegas once attracted trampling mammoths, perhaps some of our notions of what is fragile or delicate in the environment are misguided.

Although the Pinaleños have yielded no Pleistocene records of mammoths, an arthritic Columbian mammoth with fir needles in its paunch left a nearly complete set of bones at 2,700 m near the crest of the Wasatch Plateau in central Utah (Gillette and Madsen 1993). A radiocarbon date on one of the fir needles revealed that the animal died 11,420 ± 110 years before the present (B.P.). The date for hydroxyproline, an amino acid, from the bones of the mammoth was 11,220 years B.P. The dates are close to the age of the San Pedro Clovis-mammoth sites. Given the mountain-ranging habits of African elephants (G. Haynes 1991), I am confident that at least in summer the higher elevations of the Pinaleños, like those of the Wasatch, attracted Columbian mammoth and other megafauna. In Utah and Arizona the high-elevation mammoths would likely have associated with various montane small mammals such as shrews, voles, marmots, pika, golden-mantled ground squirrels, and red squirrels, all about to adjust to the Holocene warming of the climate and the arrival of human foragers. Whatever forced the extinctions of mammoths and other megafauna about 11,000 years ago, it was an environmental wreck that utterly dwarfs anything that is known to have happened in the region since.

Today the surviving small mammals isolated on the higher mountains provide grist for biogeographers and problems for stewards of the sky islands. Along with or after the evolutionary extinctions of lineages of New World megafauna, populations of forest or subalpine-adapted shrews, voles, marmots, squirrels, and other "minifauna" suffered localized range reductions. These ecological extinctions are explored by Bruce Patterson (Chapter 10), who emphasizes the value of the fossil record in illuminating biogeographic inference based on modern geographical analysis, and by Robert Sullivan and Terry Yates (Chapter 12), who consider morphological and genetic evidence. Four Rocky Mountain mammals were left in the Pinaleños following the Holocene climatic warming, which surely reduced their habitat space: a shrew

(*Sorex monticolus*), a vole (*Microtus longicaudus*), a deer mouse (*Peromyscus maniculatus rufinus*), and the red squirrel (*Tamiasciurus hudsonicus grahamensis*). Although the Pinaleños are not large, harboring a montane forest estimated at 87 km^2 (Lomolino et al. 1989), the number of refugial species seems too small to saturate the montane forest habitat. As a rough guess I would expect at least twice as many small mammals. I suggest that the missing species were squirrel-size or larger and, like squirrels, diurnal in habits. This casual prediction requires some amplification.

A variety of environmental accidents at any time could have eliminated any of the 26 species that Bruce Patterson (1984) or Mark Lomolino, James Brown, and Russell Davis (1989) provide in their list of Rocky Mountain forest mammals, species to be expected in the mountains of Arizona at least in the last cold stage of the Pleistocene. Some species of austral or Madrean origin might well have gained and lost and regained a foothold on the Pinaleños during the last 13,000 years, such as the Abert's squirrel (which has been successfully introduced by the Arizona Game and Fish Department). The survivors had to endure droughts and fires (Chapter 7; Grissino-Mayer et al. 1994), although fire might have improved conditions for meadow-inhabiting species.

Fire in old-growth spruce and fir is rare and occurs historically only under conditions of extreme drought. When old-growth spruce forest burns, perhaps on a time scale of every 300 to 400 years, the effect is catastrophic (Grissino-Mayer et al. 1994). Wet periods marked by floods (Ely et al. 1993) enhance fuel buildup preceding the next drought and fire storm. Although it obviously was several thousands of years too late to drive the extinction of megafauna, altithermal drought of 4,500 to 8,000 years ago is envisioned as severe (Waters 1989) and may have eliminated some of the isolated mammals. However, not all extinctions are driven by natural causes.

On oceanic islands, for example, investigators have recently recovered vertebrate fossils in abundance. Paleontological and archaeological efforts on islands of the West Indies (Pregill et al. 1995) and in the Pacific (Olson and James 1984; James et al. 1987; Steadman 1989, 1993, 1995) have revealed an extraordinary record of extinction. The losses on Pacific islands include flightless species of rails, geese, and ibis, as well as volant parrots, pigeons and doves, and sea birds. Endemic mammals, reptiles, and birds have also vanished from the West Indies. Extinction occurred within the last few thousand years. Prehistoric human colonization, accompanied by the introduction of pigs, dogs, and chickens (with the accidental introduction of alien species such as

Polynesian rats and land snails), reduced native bird species by 50–90% and eliminated many other native species. Conservationists and biogeographers alike are scrambling to assess the magnitude of this prehistoric human-caused wreckage.

Equally remarkable, although less thoroughly documented, is the virtual lack of faunal turnover preceding the mass extinctions accompanying human arrivals. Reduction in land mass did not trigger bird extinctions on Maui (James and Stafford 1994). Similarly, the extreme changes from wet to dry climates, for which the coastal waters of Peru and the Galapagos are famous, left no appreciable record of faunal turnover before humans arrived in the Galapagos (Steadman et al. 1991). Virtually all extinctions, two orders of magnitude above those that occurred in the Holocene, are confined in the Galapagos colonization to historic time, that is, the last few hundred years. In Tonga (Steadman 1993), Hawaii (James 1987), and New Zealand (Worthy and Holdaway 1993) alike, native fauna that were severely reduced or lost by anthropogenic (human-caused) impacts during the last few thousand years had endured tens of thousands of years of environmental change, including the late glacial warmup of the late Quaternary that radically altered sea level, ice mass, and vegetation patterns globally, without significant extinctions of island species.

What I am leading up to is a revision of ideas about the extinctions on sky islands based on the record of oceanic islands. On oceanic islands extinctions removed both taxa and populations following prehistoric human colonization. On the mountain islands of the Gadsden Purchase, by contrast, biogeographers typically consider natural causes to be responsible for the current vicariance. They have not considered prehistoric human impacts.

A variety of reasons have contributed to this state of affairs in biogeographic analysis. Foremost may be the traditional preoccupation of biogeographers with modern distributions and modern biotic communities; the fossil record shows these to be quite mutable. Another is the scarcity of well-dated archaeofaunal deposits in the western United States, although zooarchaeologists are rapidly attacking this problem when opportunities allow, as in southern Arizona (Szuter 1991) and in the Great Basin (Grayson 1993). Yet another is the likelihood that oceanic island biotas were more vulnerable to prehistoric human impacts than were those on the mainland, which lacked flightless birds and predator-naive reptiles and mammals.

All of these reasons are valid, and it would be absurd to expect the local depletions of small mammals in the mountains of North America to match anything like that seen over the last 2,000 years in islands of the eastern Pacific. By the same token, the view that Native Americans were not keystone predators, that they were "ecologically noble savages" practicing conservation is equally absurd (Alvard 1993; Kay 1994). Very likely, prehistoric foragers ranged over the mountain islands for game as the archaeological record indicates they did for timbers for midvalley habitations (Fish et al. 1992). Evidence of prehistoric hunting in high mountains is reported in northern New Mexico close to the Colorado border; archaeologists excavated 1,100 individual bones of nonhuman vertebrate animals along the San Juan River (Harris 1963). The sites occur within or near pinyon-juniper woodland, sagebrush, and grassland. Lagomorphs (hares and rabbits) predominated as they do in most archaeofaunal accounts (Szuter 1991).

What is extraordinary in Art Harris's analysis are the records of montane forest mammals that excavators found in human sites well below the forest. In a Piedra Phase pit house of the Rosa site (dated at around 900 A.D.), Harris reported remains of a red squirrel; in another the bone of a pine marten. In refuse of the early Arboles Phase (950–1000 A.D.) of the Bancos site, 6.5 km south of the Rosa site, Harris recovered bones of both red squirrel and two femurs and one tibia of a pika, which "surely represent trade items or carry-ins by the Indians" (Harris 1963:47). The possibility that the bones had been introduced by another predator is remote, since the distance to the nearest pika and red squirrel habitat "must have been no less than twenty-five miles" (Harris 1963:29). Harris had to assume that the montane or subalpine species had been brought to the sites by the human occupants. However, because of their small size it was unlikely that the animals had been brought for food. Possibly they were mummified carcasses, part of the medicine kit of a shaman. Possibly some were kept as pets. Or "perhaps the powerful force of human curiosity accounts for the presence of this animal [pika] so far from its native habitat" (Harris 1963:29).

Marmot remains in the Navajo Reservoir sites may represent food items. Harris reported 18 individual yellow-bellied marmots, a species typically associated with mountain meadows, in spruce-fir forest at 3,000 m (Harris 1963). In San Juan County, New Mexico, marmots descend to 1,800 m in pinyon-juniper habitat near Sambrito Creek (Findley et al. 1975). Marmot bones were also identified at Gila Pueblo (McKusick 1986), a late prehistoric site in Catron County, New Mexico, roughly 300 km south of the known range of the species (and within 160 km of the Pinaleños). Likewise, marmots are found in archaeological sites in the Great Basin beyond their modern range (Grayson 1993). Did marmots range more widely in the Holocene, or did prehistoric foragers travel some distance to hunt them, or do both explanations hold?

To be sure, discovering the incorporation of marmot bones in an archaeological site is far from demonstrating elimination of marmots or other populations of montane mammals by humans. In some cases the bones might have been carried from a distance. Nevertheless, paleoecological study leads to another observation regarding the curious occurrences of marmot bones in the fossil record.

In the last 40 years the Desert Laboratory on Tumamoc Hill outside of Tucson, initiated by the Carnegie Institution of Washington, has become the base of operations for a small cadre of University of Arizona geosciences students interested in the biota of the late Quaternary. They focused especially on the last 40,000 years, the interval datable by radiocarbon. These investigators screen-washed fossil packrat middens and dated over 1,000 middens by the radiocarbon method: the majority were at least 8,000 years old. The records disclosed a variety of shifts in the range of desert plants and animals, often (but not always) from lower to higher latitudes or altitudes, as one might expect given a warmer climate following the Wisconsin cold stage ending some 13,000–10,000 years ago (Betancourt et al. 1990). In some cases the radiocarbon-dated records of middens with their assemblages of plants gathered by pack rats could be aligned with fossil records of small mammals, as in the following example.

As implied above, the Rocky Mountain species that moved farthest south in the late Quaternary and left fewest isolated populations is the yellow-bellied marmot. Its fossil record includes southern Arizona and New Mexico (figure 10.1a), typically down to a minimum elevation of 1,200 m. However, in the western end of the Grand Canyon the marmot evidently descended even further, below 600 m, where marmot bones have been found in Rampart Cave, Muav Cave, and Vulture Cave, some associated with radiocarbon dates of 16,000 years B.P. (Mead and Phillips 1981). Packrat middens of that age indicate that Grand Canyon marmots lived in a dry woodland of juniper, single-leaf ash, shadscale, and prickly pear (Spaulding 1990). This arid habitat is a far cry from meadows in spruce-fir forest or alpine tundra, the habitat where marmots are supposed to thrive, even stretched to include the riparian record at 1,800 m in northern New Mexico mentioned by Harris (1963).

The problem of the marmots and their distribution is neatly illustrated by Patterson in his chapter of this book. He visually pairs the modern and fossil range of marmots with that of shrews (figure 10.1a, b). Whereas both species ranged farther south in the late Quaternary than they do now, the marmot abandoned all the mountain-island habitats retained by the shrews. The absence of marmots is intriguing. The range of the missing marmots encompasses all of Arizona and mountains in central and southern New Mexico. This area is within the heartland of the Anasazi and other prehistoric Pueblo cultures, which in prehistoric time was one of the most densely inhabited portions of the western interior of the United States; moreover, marmot remains occur in archaeological sites, as we have seen. I propose that the absence of marmots from high-mountain meadows in Arizona, including those of the Pinaleños, is a cultural, not a climatic, legacy. It follows that the modern range of yellow-bellied marmots does not encompass their potential natural distribution.

If prehistoric human foragers were intense users of the desert mountains, their ancient activities may account for many mammalian extirpations, including those of marmots and other sciurids. The absence of Abert's squirrels from isolated mountains supporting conifers below the Mogollon Rim and into the Madrean archipelago encouraged the Arizona Game and Fish Department to experiment with a series of transplants. Biologists with Game and Fish introduced fewer than one hundred Abert's squirrels on each of five mountain islands, including the Santa Catalinas and the Pinaleños. Not only did the transplants "take," yielding new populations that have sustained hunting, but the animals expanded their range by crossing "unsuitable" habitats of encinal or chaparral to invade other pine forests that previously lacked tree squirrels (Brown 1984). The success of the project suggests a "return of the native."

Abert's squirrels, like marmots, occur in the archaeological record (e.g., Harris 1963; McKusick 1986), most dramatically as a pelt (from the nose to the base of the tail) supporting feather ropes incorporating 2,336 feathers of the scarlet macaw (*National Geographic* 1982:573). Archaeologists recovered the artifact (an Indian "skirt") from a small cave in southeastern Utah (Hargrave 1979). Ceremonial use of both Abert's and other squirrels such as the Mt. Graham red squirrel is likely.

Pine martens, pikas, and other medium-sized mammals gathered by the Anasazi prehistorically (Harris 1963), could have suffered excessive predation on mountain islands if prehistoric foragers had applied effective techniques of capture, the sort of techniques seen on Pacific and other oceanic islands (Steadman 1989). Subsidized by plant foods and motivated by the chase, foraging bands of all ages would have ranged into the mountains. At Gatecliff rockshelter in central Nevada, 55 pika specimens were deposited between 7,000 and 5,000 years B.P. (Grayson 1993) — only 3 afterward. Pikas are locally absent now. Was pika extirpation driven by change in the environment or by the activity of foragers?

One of the redemptive aspects of oceanic island biogeography as developed by the James-Olson-Stafford-Steadman school is that it enables managers to consider new approaches to restoration ecology. Various extirpated bird species surviving on only one or a few islands in the Pacific may be returned to the remote islands where fossils show they once lived. If the theme of overhunting explored here has merit on the sky islands, it may open up new thoughts of where to look for more fossils and of where to reestablish populations of marmots, pikas, or red squirrels. Whether driven by climate, cultural activity, or chance, the mountain-island extinctions of the Holocene can be reversed, at least for those gene pools that survived prehistoric accidents, both climatic and cultural. Not all the genetic legacy of the Mt. Graham red squirrel need be sequestered forever in the Pinaleños. To extend Hoffmeister's thought mentioned earlier about the former range of the species, the Chiricahuas, the Santa Catalina Mountains, and other mountain islands might serve as backup reserves for a distinctive race that over the last 40,000 years miraculously survived turbulent times.

Acknowledgments

For guidance to literature and ideas, for sharing ideas, or for editorial aid I thank Scott Anderson, Conrad Bahre, Julio Betancourt, Owen Davis, Russell Davis, Gloria Fenner, George Ferguson, Paul Fish, Suzanne Fish, Patricia Gilman, Henri Grissino-Mayer, Arthur Harris, William Hartman, Charmion McKusick, Steve McLaughlin, Mary Ellen Morbeck, Wade Sherbrook, Jennifer Shopland, Patricia Spoerl, David Steadman, Mary Stiner, Peter Strittmatter, Thomas Swetnam, Peter Warshall, Anne I. Woosley, and Paul Young. For word processing I thank Jo Ann Overs and the University of Arizona Department of Geosciences. Conrad Istock and, indirectly, Don Grayson inspired this essay.

References

Alvard, M. S. 1993. Testing the "ecologically noble savage" hypothesis: Interspecific prey choice by Piro hunters of Amazonian Peru. *Human Ecology* 21:355–87.

Anderson, R. S. 1993. A 35,000 year vegetation and climate history from Potato Lake, Mogollon Rim, Arizona. *Quaternary Research* 40:351–59.

Anderson, R. S., and D. S. Shafer. 1991. Holocene biogeography of spruce-fir forests in southeastern Arizona—Implications for the endangered Mt. Graham red squirrel. *Madroño* 38:287–95.

Anderson, R. S., and S. J. Smith. 1993. *Vegetation Changes Within the Subalpine and Mixed Conifer Forests of Mt. Graham, Arizona: Proxy Evidence for the Habitat of the Mt. Graham Red Squirrel* (Tamiasciurus hudsonicus grahamensis). Final report to U.S. Department of the Interior, Fish and Wildlife Service, Phoenix, Ariz.

Bahre, C. J. 1991. *A Legacy of Change.* University of Arizona Press, Tucson.

Bennett, P. S., and M. R. Kunzmann. 1992. Terrain roughness, a feature needed to estimate plant species richness in the Madrean biogeographic region. Presented to the Fifth U.S.-Mexico Border State Conference on Recreation, Parks, and Wildlife, Las Cruces, N.M. (Unpublished.)

Betancourt, J., T. Van Devender, and P. S. Martin. 1990. *Packrat Middens: The Last 40,000 Years of Biotic Change.* University of Arizona Press, Tucson.

Bolton, H. E. 1936. *Rim of Christendom.* Macmillan, New York.

Brown, D. E. 1984. *Arizona's Tree Squirrels.* Arizona Game and Fish Department, Phoenix.

Cordell, L. 1984. *Prehistory of the Southwest.* Academic Press, Orlando, Fla.

Ely, L. L., Y. Enzel, V. R. Baker, and D. R. Cayan. 1993. A 5000-year record of extreme floods and climate change in the southwestern United States. *Science* 262:410–12.

Findley, J. S., A. H. Harris, D. E. Wilson, and C. Jones. 1975. *Mammals of New Mexico.* University of New Mexico Press, Albuquerque.

Fish, P., S. Fish, C. Brennan, D. Gann, and J. Bayman. 1992. Marana: Configuration of an Early Classic Period Hohokam platform mound site. In *Proceedings of the Second Salado Conference, Globe, Arizona, 1992,* ed. R. Lange and S. Germick, pp. 61–68. Arizona Archaeological Society Occasional Paper. Phoenix.

Gehlbach, F. R. 1993. *Mountain Islands and Desert Seas: A Natural History of the U.S.-Mexican Borderlands.* Texas A & M University Press, College Station.

Gillette, D. D., and D. B. Madsen. 1993. The Columbian mammoth, *Mammuthus columbi,* from the Wasatch Mountains of central Utah. *Journal of Paleontology* 87:669–80.

Grayson, D. K. 1993. *The Desert's Past: A Natural Prehistory of the Great Basin.* Smithsonian Institution Press, Washington, D.C.

Grissino-Mayer, H. D., C. H. Baisan, and T. W. Swetnam. 1994. *Fire History and Age Structure Analyses in the Mixed-Conifer and Spruce-Fir Forests of Mount Graham: A Final Report Submitted to the Mount Graham Red Squirrel Study Committee.* Laboratory of Tree-Ring Research, University of Arizona, Tucson.

Hargrave, L. L. 1979. A macaw feather artifact from Southeastern Utah. *Southwestern Lore* 45:1–6.

Harris, A. H. 1963. *Vertebrate Remains and Past Environmental Reconstruction in the Navajo Reservoir District.* Papers in Anthropology 11. Museum of New Mexico Press, Santa Fe.

Haynes, C. V., Jr. 1991. Geoarchaeological and paleohydrological evidence for a Clovis-age drought in North America and its bearing on extinction. *Quaternary Research* 35:438–50.

Haynes, C. V., Jr. 1992. Contributions of radiocarbon dating to the geochronology of the peopling of the New World. In *Radiocarbon Dating After Four Decades,* ed. R. E. Taylor, A. Long, and R. S. Kra, pp. 355–74. Springer-Verlag, New York.

Haynes, G. 1991. *Mammoths, Mastodonts and Elephants.* Cambridge University Press, New York.

Hoffmeister, D. F. 1986. *Mammals of Arizona.* University of Arizona Press, Tucson, and Arizona Department of Game and Fish, Phoenix.

James, H. F. 1987. A late Pleistocene avifauna from the island of Oahu, Hawaiian Islands. *Documents of the Laboratory of Geology, Faculty of Science* [Lyon] 99:221–30.

James, H. F., and T. W. Stafford Jr. 1994. 9000 years of natural vs. anthropogenic extinction in Hawaiian vertebrates. (Unpublished.)

James, H. F., T. W. Stafford Jr., D. W. Steadman, S. L. Olson, P. S. Martin, A. J. T. Jull, and P. C. Mccoy. 1987. Radiocarbon dates on bones of extinct birds from Hawaii. *Proceedings of the National Academy of Science* 84: 2350–54.

Kay, C. 1994. Aboriginal overkill: The role of Native Americans in structuring western ecosystems. *Human Nature* 5:359–98.

Lee, B. G. 1987. A deposit of stemless points from a mountain shrine. In *Mogollon Variability,* ed. C. Benson and S. Upham, pp. 219–27. New Mexico State University Occasional Papers 15. Las Cruces.

Lindsay, E. H., and N. T. Tessman. 1974. Cenozoic vertebrate fossil localities in Arizona. *Journal of the Arizona Academy of Science* 9:3–24.

Lomolino, M. V., J. H. Brown, and R. Davis. 1989. Island biogeography of montane forest mammals in the American Southwest. *Ecology* 70:180–94.

McDonald, J. A. 1993. Phytogeography and history of the alpine-subalpine flora of northeastern Mexico. In *Biological Diversity of Mexico,* ed. T. P. Ramamoorthy, R. Bye, R. A. Lot, and J. Fa, pp. 681–99. Oxford University Press, New York.

McKusick, C. R. 1986. Faunal remains. In *The Archeology of Gila Cliff Dwellings,* ed. K. M. Anderson, G. J. Fenner, D. P. Morris, G. A. Teague, and C. McKusick, pp. 245–72. Publications in Anthropology 36. Western Archeological and Conservation Center, Tucson, Ariz.

McLaughlin, S. P. 1993a. Additions to the flora of the Pinaleño Mountains, Arizona. *Journal of the Arizona-Nevada Academy of Science* 27:5–32.

McLaughlin, S. P. 1993b. Notes on the botany of the "sky islands" region of southeastern Arizona. (Unpublished.)

Marshall, J. T., Jr. 1957. Birds of Pine-Oak Woodland in Southern Arizona and Adjacent Mexico. *Pacific Coast Avifauna* 32.

Martin, P. S. 1963. *The Last 10,000 Years: A Fossil Pollen Record of the American Southwest.* University of Arizona Press, Tucson.

Martin, P. S. 1965. Geochronology of Pluvial Lake Cochise, southern Arizona, part 2, Pollen analysis of a 42 meter core. *Ecology* 44:436–44.

Martin, P. S., and R. G. Klein, eds. 1984. *Quaternary Extinctions: A Prehistoric Revolution.* University of Arizona Press, Tucson.

Mead, J. I., and A. M. Phillips III. 1981. The late Pleistocene and Holocene fauna and flora of Vulture Cave, Grand Canyon, Arizona. *Southwestern Naturalist* 26:257–88.

Meinzer, O. E., and F. C. Kelton. 1913. *Geology and Water Resources of the Sulphur Springs Valley, Arizona.* U.S. Geological Survey Water Supply Paper 320. GPO, Washington, D.C.

National Geographic. 1982. 162(5):573.

Olson, S. L., and H. F. James. 1984. The role of Polynesians in the extinction of the avifauna of the Hawaiian Islands. In *Quaternary Extinctions: A Prehistoric Revolution,* ed. P. S. Martin and R. G. Klein, pp. 768–80. University of Arizona Press, Tucson.

Patterson, B. D. 1984. Mammalian extinction and biogeography in the southern Rocky Mountains. In *Extinctions,* ed. M. H. Nitecki, pp. 247–93. University of Chicago Press, Chicago.

Pregill, G. K., D. W. Steadman, and D. R. Watters. 1995. *Late Quaternary Vertebrate Faunas of the Lesser Antilles: Historical Components of Caribbean Biogeography.* Bulletin of the Carnegie Museum of Natural History. Pittsburgh, Penn. (In press.)

Porter, S. C. 1989. Some geological implications of average Quaternary conditions. *Quaternary Research* 32:245–61.

Sellers, W. D., R. H. Hill, and M. Sanderson-Rae. 1985. *Arizona Climate: The First Hundred Years.* University of Arizona Institute of Atmospheric Physics, Tucson.

Shreve, F. 1915. *The Vegetation of a Desert Mountain Range as Conditioned by Climatic Factors.* Carnegie Institution of Washington Publication 217. Washington, D.C.

Shreve, F. 1919. A comparison of the vegetational features of two desert mountain ranges. *Plant World* 27:291–307.

Spaulding, W. G. 1990. Vegetational and climatic development of the Mojave Desert: The last glacial maximum to the present. In *Packrat Middens: The Last 40,000 Years of Biotic Change,* ed. J. L. Betancourt, T. R. Van Devender, and P. S. Martin, pp. 166–99. University of Arizona Press, Tucson.

Spicer, E., II. 1962. *Cycles of Conquest: The Impact of Spain, Mexico, and the United States on the Indians of the Southwest.* University of Arizona Press, Tucson.

Spoerl, P. M. 1992. Shrine sites on Mt. Graham: Their significance and to whom. Paper presented at the Seventh Mogollon Conference, 25–26 September 1992, Las Cruces, N.M.

Steadman, D. W. 1989. Extinction of birds in eastern Polynesia: A review of the record and comparisons with other Pacific Island groups. *Journal of Archaeological Science* 16:177–205.

Steadman, D. W. 1993. Biogeography of Tongan birds before and after human impact. *Proceedings of the National Academy of Science* 90:818–22.

Steadman, D. W. 1995. Prehistoric extinctions of Pacific island birds: Biodiversity meets zooarchaeology. *Science* 267: 1123–31.

Steadman, D. W., T. W. Stafford Jr., D. J. Donahue, and A. J. T. Jull. 1991. Chronology of Holocene vertebrate extinction in the Galapagos Islands. *Quaternary Research* 36:126–33.

Szuter, C. R. 1991. *Hunting by Prehistoric Horticulturists in the American Southwest.* Garland Publishing Co., New York.

Taylor, R. J., T. F. Patterson, and R. J. Harrod. 1994. Systematics of Mexican spruce—revisited. *Systematic Botany* 19:47–59.

Warshall, P. 1986. *Biogeography of the High Peaks of the Pinaleños.* Maricopa Audubon Society, Phoenix, Ariz. Reprinted from *The Environmental Data Book,* U.S. Forest Service, Coronado National Forest.

Waters, M. R. 1989. Late Quaternary lacustrine history and paleoclimatic significance of Pluvial Lake Cochise, southeastern Arizona. *Quaternary Research* 32:1–11.

Waters, M. R., and A. I. Woosley. 1990. The geoarchaeology and preceramic prehistory of the Willcox Basin, southeastern Arizona. *Journal of Field Archaeology* 17:163–75.

Whittaker, R. H., and W. A. Niering. 1964. Vegetation of the Santa Catalina Mountains, Arizona, part 1, Ecological classification and distribution of species. *Journal of the Arizona Academy of Science* 3:9–34.

Whittaker, R. H., and W. A. Niering. 1965. Vegetation of the Santa Catalina

Mountains, Arizona, part 2, A gradient analysis of the south slope. *Ecology* 46:429–52.

Whittaker, R. H., and W. A. Niering. 1975. Vegetation of the Santa Catalina Mountains, Arizona, part 5, Biomass production and diversity along the elevation gradient. *Ecology* 56:771–90.

Worthy, T. H., and R. N. Holdaway. 1993. Quaternary fossil faunas from caves in the Puna Kaiki area, west coast, South Island, New Zealand. *Journal of the Royal Society of New Zealand* 23:147–254.

Epilogue

We are becoming poor in wilderness faster than most of us know. The speed of loss is great, there are few remote places, and they will not multiply or fuse or become a frontier again.

Facing today's kind of frontier, we see a challenge that will engage the best research, education, human creativity, and drive we can produce. The new challenge is to retrace steps through mistreated land, learn to heal, to rebuild, to reinvigorate it—and pay whatever this new attitude costs us. We must increasingly substitute resourcefulness for resources in order not to overdraw an account our children would have to cover for us.

DAVID BROWER, *For Earth's Sake*

Always I am aware of maintaining a harmonious, natural-cultural mosaic throughout the Borderlands, a landscape diversity managed by the controlled growth of culture. Alternatively, I visualize an unending, man-made desert on border trips of the twenty-first century. Whereas those intrepid surveyors and naturalists of the nineteenth century gave man his first glimpse of the Border's diverse physical and biotic resources and twentieth-century naturalists have documented what fraction remains and how it harmonizes in an arid context, travelers henceforth must be concerned primarily with the interplay of resource keeping and using—the rapprochement of man and nature. The most difficult stewardship is yet to come.

FREDERICK R. GEHLBACH, *Mountain Islands and Desert Seas*

The Mt. Graham affair has raised issues far larger than specific regional conflicts over astrophysical development. Not only did the controversy spill over into state offices, the national media, and Congress, it also called into question the adequacy of the laws, procedures, and institutions meant to protect the natural resources of the United States; it exposed levels of social and scientific complexity not usually acknowledged in discussions of modern environmentalism; it released a mixture of human foibles and confusion fit for tragic theater; it brought us face to face with the reality of the damaged past, enduring splendor, and uncertain future of a Southwestern mountain island—perhaps bringing the Pinaleños into sharper biological focus than ever could have occurred without the societal storm that gathered over them. The rugged mountain fastness of the Pinaleños has become an exemplar for all the other mountain islands of this extraordinary inland archipelago and, we hope, will provide a strong stimulus for the preservation of biological diversity all across our country. If we can make the knowing and caring development of a scientifically valuable astrophysical observatory harmonious with such a grand part of nature, perhaps we can go on to gather the knowledge and wisdom to perpetuate such harmony elsewhere and indefinitely into the future. The Mt. Graham experience may have taught us something about how we can approach the larger and more difficult tasks of stewardship over the damaged lands and waters that now lie from coast to coast across America.

Perhaps, though, we had best take first things first. We must ensure that the restoration and preservation of the wildness of the Pinaleños is permanent policy. Then we must make real the suggestion that a mountain island preserve be established to guarantee that the remaining biota of the archipelago will endure to the greatest possible extent. This would be a major departure from the limitations of weaker attempts at preservation based on the National Environmental Policy Act and the Endangered Species Act. We must do the scientific studies necessary to make the stewardship of wild lands more sensitive and effective than it has ever been in the past. Let the Arizona congressional delegation, which addressed the stubborn Mt. Graham conflict with an amendment passed in Congress, now formulate and propose the "Mountain Island Conservation Act." If this much can be done, a great deal more can follow.

How might we avoid such costly battles? We need a triggering mechanism that cries "Stop" when an environmental conflict of potentially large magnitude first arises. It must be followed by an objective assessment conducted by noncombatants. This certainly means that people outside the responsible federal and state agencies should be consulted at the outset. The new process must include a prescription for scientific data gathering and analysis and a determination of when information is sufficient to guide us to a decision. That accomplished, the battle may rage anyway, but we have reason to hope that scientific objectivity early on will invite calm and eventually wise decisions. It is clear that present practices under the National Environmental Policy Act and the Endangered Species Act are typically not so perspicacious. However, installing objectivity in the ways we go about implementing these laws may be more productive than a lot of effort expended revising them.

Pondering the sweep of this book—including the sense of deep history to which Paul Martin draws attention — and coming to the profound present-day challenges posed in the above quotes from Brower and Gehlbach, we recognize that the struggle to achieve what they propose is now a desperate struggle. Failure will make a wasteland of our continent; success will leave a great legacy for those who follow us, and it will be a national triumph for "these United States."

Contributors

About the Editors

Conrad A. Istock received a Ph.D. in zoology at the University of Michigan with an emphasis on ecology and evolutionary biology. He taught for many years in the Department of Biology at the University of Rochester. He is currently a professor in the Department of Ecology and Evolutionary Biology at the University of Arizona. He has published papers on ecology, population genetics, and evolution. His studies have involved field and laboratory experimentation and mathematical theory. His research has included studies of insect life history, genetics, and evolution; landscape-scale analysis of plant communities; life history theory; and the genetics and ecology of bacterial populations.

Robert S. Hoffmann is the provost of the Smithsonian Institution. He is a world authority on the evolution of Holarctic mammals, as well as the mammals of the former Soviet Union, China, and central Asia. He earned his M.S. and Ph.D. from the University of California, Berkeley. He held academic positions at the Universities of Montana and Kansas before joining the Smithsonian as director of the National Museum of Natural History. His career in research and administration has given him extensive experience in the international scientific arena, as well as in the field of science policy.

About the Contributors

Russell Davis's research interests have centered around Southwestern biogeography and the natural history of Southwestern mammals. As a montane field ecologist he has great appreciation for the emotional importance of southeastern Arizona's isolated mountain islands—as special and mysterious places for exploration and adventure—as well as an understanding of their significance as centers for biogeographical, ecological, and evolutionary research. He earned his Ph.D. from the University of Arizona and returned for a career of teaching and research.

Harold C. Fritts is Professor Emeritus of the Laboratory of Tree-Ring Research, University of Arizona; adjunct professor at the Biological Sciences Center, Desert Research Institute, University of Nevada; and owner of Dendro-Power, a dendroecological modeling and consulting business. Simulation and empirical modeling of tree-ring structure as it is affected by environmental factors is his primary interest at present. The primary topics of his research have been synoptic scale climatic reconstruction from tree-ring grids, dendroecology, tree-growth environmental relationships, and developing and managing the International Tree-Ring Data Bank.

Frederick R. Gehlbach is author of *Mountain Islands and Desert Seas: A Natural History of the U.S.-Mexican Borderlands,* now in its second edition. Much of his research has focused on the life history, ecology, and behavior of owls. He has served as ecological consultant to a large number of governmental agencies and nongovernmental organizations throughout his career. He earned his Ph.D. at the University of Michigan and is a professor of biology at Baylor University.

Henri D. Grissino-Mayer is interested in long-term dendroclimatic reconstructions, the fire ecology and history of Southwestern mixed-conifer and ponderosa pine forests, and the relationship between paleoclimate and past fire occurrence. He has been conducting dendroclimatic and fire history research in the Mt. Graham area since 1988 and is currently expanding this research to the Chiricahua Mountains south of the Pinaleño Mountains. He serves as forum manager for the International Tree-Ring Data Bank located at the National Geophysical Data Center in Boulder, Colorado, and for the Biogeography Specialty Group of the Association of American Geographers. He is currently a graduate research associate in the Laboratory of Tree-Ring Research at the University of Arizona.

Martin Harwit is the former director of the National Air and Space Museum of the Smithsonian Institution. Before joining the Smithsonian, he held a number of academic positions at Cornell University and the Max Planck Institute for Radioastronomy, Bonn, West Germany. He is the author of three books and has been involved in the production of documentaries for public television and film. He received his Ph.D. in physics from MIT in 1960.

Lawrence R. Heaney developed his early interest in biology, especially biogeography and systematics, at the Smithsonian Institution. He is currently associate curator and head of the Division of Mammals at the Field Museum in Chicago. His primary research interest is in patterns of biological diversity, including both their evolutionary origin and their ecological maintenance. He is also heavily involved in providing advanced training to young professional conservation biologists from the tropics. He received his M.S. and Ph.D. from the University of Kansas.

Paul S. Martin is Emeritus Professor of Geosciences at the University of Arizona. He has spent his entire professional career with University of Arizona paleoecology students at the Desert Laboratory on Tumamoc Hill on the west side of Tucson. Besides explorations in Africa, Australia, Argentina, and Mexico, he has traveled, figuratively speaking, into the late Quaternary via fossil deposits in the Grand Canyon and other parts of the Southwest, hunting for the lost ground sloths and other extinct megafauna from 11,000 years ago. He is best known for his controversial views on overkill by prehistoric humans in the New World.

Duncan T. Patten is a professor of botany and the director of the Center for Environmental Studies at Arizona State University. He has studied the plant ecology of ecosystems of the Southwest, including the Sonoran Desert, interior chaparral, riparian ecosystems, and conifer forests. He was the business manager of the Ecological Society for 15 years, has served on numerous National Academy of Science committees, and is presently senior scientist for the Glen Canyon Environmental Studies. His research interests emphasize the consequences of human activities on semiarid ecosystems.

Bruce D. Patterson is curator of mammals at the Field Museum in Chicago and serves on the faculties of the University of Chicago, the University of Illinois at Chicago, Northern Illinois University, and the Universidad de San Marcos in Lima, Peru. He received his M.S. and Ph.D. from New Mexico State University. His studies of biological diversification have carried him to Chile, Peru, and various parts of Brazil, focusing on such groups as opossums, bats, and especially rodents. However, his love for the Southwest and his enduring fascination with its wildlife are reflected in a continuing series of articles dealing with the distribution and extinction of relict species.

Ross Simons is currently assistant provost for science at the Smithsonian Institution. His research interests include natural resource economics, science policy, the development of field research stations, and international science and development issues. Mr. Simons is trained in history, international relations, and economic development. He serves on a number of international boards and commissions concerned with science.

Christopher Smith received his M.S. and Ph.D. at the University of Washington. His dissertation work on pine squirrel ecology and evolution has developed into a career-long interest in the evolution of arboreal mammals and tree reproduction. An interest in plant reproduction led to a paper with Steve Fretwell on the trade-off between size and number of offspring that was recognized as a "Citation Classic." He has been in the Biology Division at Kansas State University for most of his career.

Peter A. Strittmatter has studied rotation and magnetism in stars, properties of white dwarfs and other degenerate stellar configurations, the nature of quasars and related active galactic nuclei, methods of image reconstruction, and telescope design. He has been involved in the Mt. Graham International Observatory project since 1981, as director of the University of Arizona's Steward Observatory in Tucson. He earned his B.A., M.A., and Ph.D. at Cambridge University, England, in applied mathematics/theoretical astrophysics.

Juliet C. Stromberg is an associate professor of research at the Center for Environmental Studies at Arizona State University. Threatened and endangered ecosystems of the American Southwest, including old-growth forests and riparian (streamside) forests, have been the primary sites for her research. Her primary topics of research have included forest community dynamics and autecology and reproductive ecology of dominant tree species, with the intent of determining ecological needs for long-term ecosystem sustainability.

Robert Miles Sullivan is senior biologist at the Physical Science Laboratory, Environmental and Atmospheric Sciences Division, New Mexico State University. His research interests include the theory and methodology of historical and ecological biogeography, gene-flow models, conservation and endangered species biology, population genetics, and systematic biology, particularly on relict populations of Southwest coniferous forest and Great Basin mammals. Extensive field research has taken him throughout North America and western Canada, especially the American Southwest, Great Basin, and Texas. He received his Ph.D. in biology at the University of New Mexico.

M. Mitchell Waldrop is the author of *Man-Made Minds* (1987), a book about artificial intelligence, and *Complexity* (1992), a book about the Santa Fe Institute and the new sciences of complexity. He is currently at work on a book about computers, to be called *The Technology of Enchantment.* From 1977 to 1980 he was a writer for *Chemical and Engineering News;* from 1980 to 1991 he served as a senior writer at *Science.* He earned his Ph.D. in elementary particle physics at the University of Wisconsin after earning a M.A. in journalism at the same institution. He lives in Washington, D.C.

Terry L. Yates conducts research on the effects of global climate change on small mammals on mountaintops, the surveillance and monitoring of the hantavirus in natural populations of mammals, and mammals of Bolivia, and is involved with the Sevilleta Long-Term Ecological Research project. He has authored many scientific papers dealing with biodiversity. He earned his Ph.D. in 1978 from Texas Tech University. He is currently director and curator of mammals, birds, and biological materials at the Museum of Southwestern Biology and is a professor of biology at the University of New Mexico.

Paul J. Young is an assistant research scientist at the University of Arizona. He has been supervisor of the University of Arizona's Mt. Graham red squirrel monitoring program since 1989. He earned a Ph.D. in zoology at the University of Alberta in 1988. His research interests include the behavioral and physiological ecology of sciurids with emphasis on winter ecology and hibernation, particularly on the ecology and behavior of the Mt. Graham red squirrel, in comparison to other red squirrel populations in the southwestern United States.

Index